Rhumb Lines and Map Wars

RHUMB LINES AND MAP WARS

A Social History of the Mercator Projection

MARK MONMONIER

The University of Chicago Press
Chicago and London

Mark Monmonier is Distinguished Professor of Geography at Syracuse University. He is the author of numerous books, including *Spying with Maps*; *Air Apparent: How Meteorologists Learned to Map, Predict, and Dramatize Weather*; and *Bushmanders and Bullwinkles: How Politicians Manipulate Electronic Maps and Census Data to Win Elections*, all published by the University of Chicago Press.

The University of Chicago Press, Chicago 60637
The University of Chicago Press, Ltd., London
© 2004 by The University of Chicago
All rights reserved. Published 2004
Printed in the United States of America

13 12 11 10 09 08 07 06 05 04   1 2 3 4 5

ISBN: 0-226-53431-6 (cloth)

Library of Congress Cataloging-in-Publication Data

Monmonier, Mark S.
    Rhumb lines and map wars : a social history of the Mercator projection /
Mark Monmonier.
    p. cm.
Includes bibliographical references and index.
ISBN 0-226-53431-6 (cloth : alk. paper)
  1. Mercator projection (Cartography)   2. Cartography—Social aspects.
3. Loxodrome.   4. Peters projection (Cartography)   5. Navigation.   I. Title.
GA115.M66 2004
526'.82—dc22

25.00   2003027614

Dedicated to the memory of Gary Turley (1952–2003),
journalist, editor, and unreconstructed basketball fan

Contents

Preface

I conceived this project in mid-June 2001, shortly after participating in a discussion of maps and propaganda on *The Public Interest,* a midday interview/talk program on National Public Radio hosted by Kojo Nnamdi. The panel included historian Susan Schulten and the National Geographic Society's chief cartographer, Allen Carroll. In addition, Ward Kaiser, former head of Friendship Press, joined the program at one point by telephone. Kaiser is a strong supporter of the Peters projection, a rectangular world map that German historian Arno Peters began promoting in the early 1970s. Although the projection severely distorts shape in the tropics and upper latitudes, it preserves relative area. Even though numerous other map projections also preserve relative area, Peters claimed that his map was uniquely "fair to all peoples" and thus a perfect antidote to the Mercator projection, which significantly diminishes the size of developing nations, mostly in the tropics. Kaiser's presence on the program indicated that the map wars of the 1970s and 1980s were hardly over.

By "map wars" I mean the often-heated attacks by Peters and his supporters on a cartographic establishment he blamed for belittling the Third World by promoting the Mercator map, or at least failing to limit its use. The news media reported Peters's claims uncritically except for interviewing noted cartographers like Arthur Robinson, who contributed balance and excitement by offering opposing views often laced with sarcasm. Postmodern scholars like the late Brian Harley entered the fray by attacking "the 'cartographers know best' fallacy" and interpreting "Peters's agenda [as] the empowerment of those nations of the world he felt had suffered a historic cartographic discrimination"—no matter that other map projections portray the less developed world more effectively and soundly. The appearance of the Peters map on NBC's *The West Wing* in early 2001 and continued efforts by diversity training firms to promote the Peters projection suggest that the map wars have entered a guerrilla phase. *Rhumb Lines and Map Wars* offers a vigorous and needed response to a campaign that mixes willful ignorance and misguided activism.

Well-intentioned people claim that educators, politicians, and the media need the ill-conceived Peters world map because the Mercator projection predominates among classroom wall maps and thus shapes most people's mental image of the world. To be sure, the Mercator map has been widely misused for world maps having little or nothing to do with navigation, but to frame the debate narrowly as a Peters-versus-Mercator contest is not only disingenuous but blatantly ignorant of an important projection's historically significant (but fading) contributions to exploration and transportation, as well as the contributions of its more recent transverse orientation, which has proved extraordinarily useful for large-scale, detailed topographic maps. While misuse of the Mercator world map is difficult to eradicate, the effects are easily exaggerated. An appropriate solution lies in a fuller understanding of map projections and their distortions, not the contorted contours of the Peters perspective.

I have made this examination of the Mercator map deliberately broad by including the transverse Mercator projection, used widely for topographic maps, military grids, and electronic geographic databases, as well as the oblique Mercator projection, valuable in aeronautics, and the Space Oblique Mercator projection, used to add meridians and parallels to satellite imagery. These extensions reflect a fuller development of conformality, a mathematical property that connotes the true representation of angles within infinitesimally small neighborhoods around all points on a map projection. Although Mercator's 1569 map of the world pioneered the use of conformality on a rectangular projection, the property was little appreciated until the mid-eighteenth century, when Johann Heinrich Lambert used calculus to derive a family of conformal projections, of which Mercator's was a special case. Although conformality cannot avoid distorting continents and other large shapes that exist properly only in three dimensions, it ensures the reliable portrayal of streets, streams, and property boundaries on detailed local maps.

This broader treatment of the Mercator map helps demonstrate a key notion in the projection's social history, namely, that innovations in cartography, as in other science-based technologies, are rarely the work of single individuals. And like other clever developments that were ahead of their time, the Mercator projection was not widely adopted until the compass and chronometer were perfected and mariners learned to understand and trust navigation charts.

My intent is to put the Mercator projection in context by demonstrating its crucial roles in navigation and topographic mapping as well as its inappropriateness for general purpose world maps. As I show, misuse of the Mercator map in atlases and textbooks fell off markedly after World War II, when enhanced awareness of air power fostered an appreciation of polar projections and other valuable but largely neglected cartographic perspectives. Particularly puzzling is the failure of many pro–Third World groups to appreciate another kind of map projection, the area cartogram, which allows dramatic, socially relevant worldviews by distorting the mapped areas of countries to represent population, wealth, food production, or military spending.

While my prime goal is informed skepticism of both the Mercator map and its critics, I hope to leave readers with a fuller appreciation of how map projection works. That said, this is not a textbook on map projection—numerous comprehensive guides are available, and there is little need for another. To avoid needless technical details, I include only two mathematical formulas, the trigonometric expansion underlying Edward Wright's pioneering projection tables and the more compact logarithmic equation Henry Bond discovered a half-century later. Like the better map projection textbooks, I rely heavily on Nicolas Auguste Tissot's indicatrix, a straightforward graphic device for describing patterns of areal and angular distortion. Better to show the diverse distortions of the Mercator map and its various substitutes than merely talk about them.

Any attempt to show how map projections work must include their rhetorical role, which involves goals markedly different from traditional cartographic tasks like describing boundaries, exploring patterns, and getting around. This rhetorical prowess, rooted at least as much in the map's symbols and generalizations as in its projection, makes the map vulnerable to diverse ideological interpretations. Thus the Mercator map can be viewed as an icon of Western imperialism while the Peters map can connote fairness and support for Third World concerns.

A word about my own biases and point of view seems in order. Hardly unsympathetic to Third World issues, I recognize the ideological value of maps, including their persuasiveness in dramatizing problems and setting agendas. Politically I am a liberal in the traditional sense of "making government work," and I also recognize the need for social programs that work and modest amounts of affirmative action. Whether I am part of the cartographic establishment depends on your point of view. An early contributor to the development of digital mapping, I've been more a critic than a methodologist in recent years (a constructive critic, I hope), and occasionally, as now, I've been a critic of other critics. This role betrays my mixed feelings about "critical" scholars who ask penetrating questions about a map's authority and legitimacy but cloud their explorations in needlessly inaccessible jargon. I resented overly complex language when I edited *The American Cartographer* (now *Cartography and Geographic Information Science*), and I don't like it any better in the writings of social scientists and humanists. Verbal language, like its graphic counterpart, can be used to confuse or clarify. In *Rhumb Lines and Map Wars,* I hope it's the latter.

Acknowledgments

This book benefited substantially from the knowledge and persever-ance of librarians at Syracuse University, Cornell University, the U.S. Library of Congress, and the National Oceanic and Atmospheric Ad-ministration's Central Library in Silver Spring, Maryland. Syracuse specialists whose help was particularly valuable include geography and maps librarian John Olson, physical geography and earth science librarian Elizabeth Wallace, and mathematics librarian Mary DeCarlo. Special thanks for impressively swift service go to Dorcas MacDonald and the interlibrary loan staff at Bird Library. I also appreciate the as-sistance of James Flatness, John Hessler, Ron Grim, and Gary Fitz-patrick in the Geography and Map Division of the U.S. Library of Congress. Robert Wilson in the Cartographic Research Division of the National Ocean Service provided valuable insights on current and past practices in nautical charting, and Robert McMaster, a geogra-pher at the University of Minnesota and an old friend, supplied the photograph of J. Paul Goode. David Woodward at the University of Wisconsin–Madison and Ingrid Kretschmer at the University of Vi-enna contributed useful suggestions and information. At Syracuse University, the Maxwell School of Citizenship and Public Affairs and the Office of the Vice Chancellor for Academic Affairs helped with travel and research expenses. Among colleagues in the geography department I am especially indebted to staff cartographer Joe Stoll, systems manager Brian Von Knoblauch, and summer graduate research assistant Karen Culcasi. John Western, our department chairperson for the past three years, provided invaluable moral support and unflagging good humor. Janet Brieaddy and Chris Chapman in the department office helped in countless ways. Tony Campbell, former map librarian at the British Library, and Robert W. Karrow Jr., curator of maps at the Newberry Library, offered valuable suggestions on early drafts of chapters covering medieval and Renaissance cartography. I also appreciate Karrow's later comments and suggestions on the entire manuscript as well as the equally valuable advice of Keith Clarke at the University of California, Santa Barbara, and Nancy Obermeyer at Indiana State University.

I am deeply indebted to several people at the University of Chicago Press. My editor, Christie Henry, supplied insight and understanding, and her assistant, Jennifer Howard, ably fielded queries while Christie was out of the office for the birth of Eleanor. It was especially gratifying to work again with freelancer Jenni Fry, my copy editor for *Spying with Maps.* Christine Schwab, production editor, Ryan Li, designer, and Siobhan Drummond, production controller, dealt skillfully with design, page layout, and production. I also appreciate the continued support, on this and other projects, of publicity director Erin Hogan and marketing director Carol Kasper, as well as the impressive efforts of Stephanie Hlywak in promoting this book. And with no regrets after more years of marriage than we care to advertise, I thank Marge for being there.

Bearings Straight — An Introduction

Mariners share two fears: bad weather and getting lost. Their deep respect for the Mercator projection reflects the map's value for plotting an easily followed course that can be marked off with a straightedge and converted to a bearing with a protractor similar to the semicircular plastic scales fourth graders use to measure angles. In a less direct way, the Mercator map also addresses the sailor's fear of storms by providing a reliable base for plotting meteorological data for tropical regions. But that's another story.

Picture yourself as a seventeenth-century navigator who knows where he is and where he wants to go. You plot both locations on a chart, join them with a straight line, and measure the angle your line makes with the map's meridians, which run due north. If the chart is a Mercator map, all its meridians are straight lines, parallel to one another, and the course you've just plotted is a rhumb line, also called a loxodrome (fig. 1.1). The derivation of *rhumb* is obscure—possible origins include a Portuguese expression for course or direction (*rumbo*) and the Greek term for parallelogram (*rhombos*)—but math-

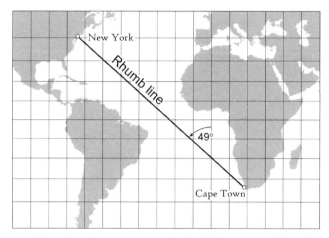

Figure 1.1 When plotted on a Mercator grid, a rhumb line intersects all meridians at the same angle. In this example a constant bearing of forty-nine degrees west of north will take a ship from Cape Town to New York.

ematician Willebrord Snell (1580–1626) coined *loxodrome* in 1624 by combining the Greek words for oblique (*loxos*) and course (*dromos*). Manuals on piloting accept *rhumb* as a normal part of the seaman's language and define *rhumb line* as a line that intersects all meridians at the same angle. The angle between a course and a meridian is a bearing, thus a rhumb line is a line of constant bearing. Stay the course, and you'll reach your destination.

Look down at a globe, on which the meridians meet at the North Pole, and you'll understand why loxodromes are spirals that converge toward the pole as they wind round and round, always crossing the meridians at a constant angle (fig. 1.2). The only exceptions are rhumb lines running directly north–south, along a meridian, or directly east–west, along a parallel. The former reach the pole along the shortest possible route, whereas the latter never get any farther north or south. If a bearing is close to due north, its loxodrome approaches the pole rapidly. If a bearing is nearly due east, convergence is notably slower, with a loxodrome that originates in the tropics and circles the globe many times before crossing the Arctic Circle. Follow a loxodrome in

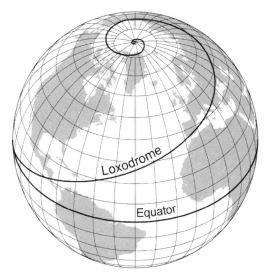

Figure 1.2 Loxodromes spiral inward toward the poles.

the other direction, and it crosses the equator and starts spiraling toward the South Pole. What works in the northern hemisphere works equally well south of the equator.

Gerard Mercator (1512–94) understood loxodromes. Skilled in engraving and mathematics, he crafted globes and scientific instruments as well as maps. Like other sixteenth-century globe makers, he engraved the grid lines, coastlines, and other features on copper plates and printed the curved surface in flat sections, called gores, which were trimmed and pasted onto a ball, typically made of papier-mâché. His first experience with globe making occurred around 1537, when he engraved the lettering for a terrestrial globe designed by his mathematics tutor, Gemma Frisius (1508–55). That same year Mercator produced his first map, a six-sheet representation of the Holy Land. In 1541, he devised a navigator's globe on which rhumb lines spiraled outward from compass roses. Intended as a navigation instrument, the globe was approximately 16.5 inches (42 cm) in diameter and consisted of the twelve gores and two polar caps pasted onto a hollow wooden ball for use at sea. According to cartographic historian Robert Karrow, this navigator's globe was the first of its kind, and sixteen surviving copies, crafted between 1541 and 1584, attest to its success and durability.

Mercator published his celebrated world map of 1569 as a set of eighteen sheets, which form a wall-size mosaic 48 inches (124 cm) tall by 80 inches (202 cm) wide. Its projection revolutionized navigation by straightening out rhumb lines on a flat map—not just the globe's meridians and parallels, but any rhumb line a seaman might plot. To accomplish this, Mercator progressively increased the separation of the parallels. On a grid with a constant separation of ten degrees between adjoining meridians and parallels—cartographers call this a ten-degree graticule—parallels near the equator are relatively close, whereas those farther poleward are more widely spaced, as shown in figure 1.3. The parallels at 70° and 80° N, for instance, are much farther apart than the equator and its neighbor at 10° N. And the separation between 80° and 90° N cannot be shown completely because the North Pole lies at infinity. Although loxodromes converge toward the poles, on a Mercator projection they never really get there.

Mercator's intent is readily apparent in his map's title, "New and More Complete Representation of the Terrestrial Globe Properly Adapted for Use in Navigation." In 1932, the *Hydrographic Review* published a literal translation of the map's numerous inscriptions, elegantly engraved in Latin. Although the chartmaker's words reveal little about how he spaced the parallels, Mercator clearly recognized the need "to spread on a plane the surface of the sphere in such a way that . . . the forms of the parts be retained, so far as is possible, such as they appear on the sphere." Accurate bearings, he reasoned, demand a locally exact representation of angles and distances, even though "the shapes of regions are necessarily very seriously stretched."

To compensate for the local deformation that would otherwise occur, Mercator "progressively increas[ed] the degrees of latitude toward each pole in proportion to the lengthening of the parallels with reference to the equator." Sounds complicated, but it's not. At 60° N, for instance, the distance on a globe between two meridians is half the

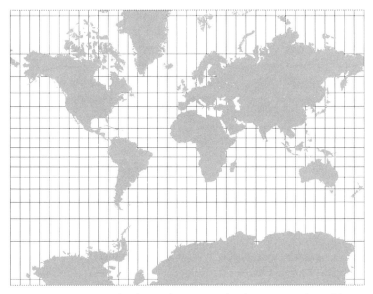

Figure 1.3 A Mercator grid with a ten-degree graticule. The map stops short of the poles because of increased north–south stretching.

corresponding distance at the equator. Because the projection doesn't let the meridians converge, it must stretch the sixtieth parallel to twice its true length. To compensate for this pronounced east–west stretching along the map's parallels, the projection imposes an identical stretching in the north–south direction, along the meridians. Farther north, as east–west stretching grows progressively larger, north–south stretching increases proportionately. At the North Pole, a mere point on the globe, map scale becomes indefinitely large—the result of stretching a dimensionless spot to a measurable distance—and the pole lies "at infinity," or at least well off the map. That's why Mercator world maps typically cut off northern Greenland and omit most of an otherwise humongous Antarctica.

Forcing north–south scale to equal east–west scale at all points not only preserves angles and bearings but prevents the deformation of small circles into ellipses. Modern cartographic textbooks consider

this locally exact portrayal of angles and small shapes, called conformality, highly desirable for detailed, large-scale maps of small areas. In
addition to depicting city blocks as rectangles, not parallelograms, a
conformal map keeps squares square and circles circular. Although
more than a century passed before Edmund Halley (1656–1742) recognized conformality as a mathematical property, Mercator's 1569
world map became the first conformal projection to portray meridians and parallels as straight lines.

In addition to drawing on his experience in making globes, Mercator borrowed a concept embedded in fourteenth-century regional
sailing charts. Portolan sailing charts, named after the *portolani,* or pilot books, that guided sailors across the Mediterranean Sea or along
the coast of Europe, were distinguished by a network of straight-line
sailing directions that converge at assorted compass roses. A typical
portolan chart was oriented to magnetic north, covered less than one-
fiftieth of the earth's surface, and lacked a consistent grid of meridians
and parallels. Originally drawn to illustrate books of written sailing directions, portolan charts reduced the uncertainty of navigating across
open waters. As the first whole-world sailing chart, Mercator's map
made a transatlantic journey look as straightforward as a voyage from
Athens to Alexandria.

However easy to navigate, a loxodrome is rarely as direct as the
great circle crudely approximated by a taunt thread stretched across a
globe between a route's origin and destination. Great circles, so called
because they are the largest circles one can draw on a sphere, define
the shortest path between two points. Although geometrically efficient, they are difficult to navigate because the bearing is constantly
changing. The only exceptions are routes along a meridian or the
equator. Because a loxodrome is not a great circle, the sailor taking its
more easily followed course takes an indirect route. But if the increased distance is long enough to make a difference, the navigator
can divide the route into sections and follow the rhumb line for each
part. In figure 1.4, a dotted line illustrates a sectioned route from Cape
Town to New York. Because the Mercator grid distorts distance, the
single rhumb line marking the constant-bearing route looks decep-

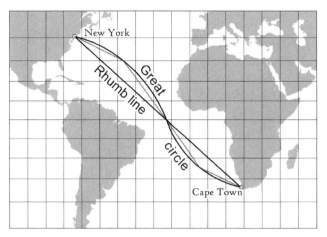

Figure 1.4 To approximate a great-circle route, which is shorter than a loxodrome, navigators often select one or more intermediate points and connect them with rhumb lines.

tively shorter than either the great-circle route or its multi-rhumb approximation.

Mercator sought to reconcile the navigator's need for a straightforward course with the trade-offs inherent in flattening a globe. These trade-offs include distortions of distance, gross shape, and area. Although all world maps distort most (if not all) distances, some projections, including Mercator's, afford negligible distortion on large-scale detailed maps of small areas. Only a globe can preserve continental outlines, however, which cannot be flattened without noticeable stretching or compression. Relative size, which is preserved on map projections with a property called equivalence, is markedly misrepresented on Mercator charts because of the increased poleward separation of parallels required to straighten out loxodromes. Distortion of area is most apparent in the chart's inflated portrayal of Greenland as an island roughly the size of South America. On a globe Greenland is not quite an eighth as large.

Like many innovations, the new projection did not catch on right away. One impediment to a wider, swifter adoption was the lack of a detailed procedure for progressively separating the parallels. Wordy inscriptions explained the map's purpose but offered no instructions for constructing or refining its grid. That Mercator produced a generally accurate solution for the lower and middle latitudes was quite an accomplishment in an era with neither logarithms to expedite calculation nor integral calculus to derive a concise mathematical formula. Trigonometric tables of secants and tangents, which might have been especially useful, were also lacking. Some scholars think Mercator used a mathematical approximation to lay out parallels ten degrees apart; a few suggest that he developed the separations graphically by copying loxodromes from a globe to a map. Whatever his approach, Mercator's map stimulated further work by English mathematicians Edward Wright (1561–1615) and Henry Bond (1600–1678), discussed in chapter 5. In 1599, in a treatise with a long title that begins *Certaine Errors in Navigation,* Wright included a table of "meridional parts," with which a chartmaker or seaman could efficiently lay down a Mercator grid. And in 1645, Bond suggested a mathematical formula after discovering a similarity between Wright's table and a table of logarithmic tangents.

Another obstacle was the primitive technology for taking compass readings at sea and correcting for magnetic declination. An inscription on the 1569 world map discusses the vexing discrepancy between the poles that anchor the earth's grid and the poles believed to attract compass needles. Eager to include a north magnetic pole on his map, Mercator consulted "a great number of testimonies," which suggested diverse positions for a magnetic meridian aimed at the magnetic pole. Some observations placed this magnetic meridian in the Cape Verde Islands, where magnetic north coincided with true north; others placed it at Corvo, in the Azores. Equally suggestive was Marco Polo's report that "in the northern parts of Bargu [in northeast Asia] there are islands, which are so far north that the Arctic pole appears to them to deviate to the southward." Without marking the Corvo meridian explicitly on his map, Mercator extended it up over the pole and then south toward Asia. In doing so, he wrongly assumed that compass needles point along great circles that converge at the magnetic poles.

Aware that, because of this uncertainty, the location didn't warrant an X or a compass rose, Mercator marked the spot with what looks like a fried egg with a very small yoke (fig. 1.5). An adjacent inscription restates the premise: "It is here that the magnetic pole lies if the meridian which passes through the Isle of Corvo be considered at the first." To hedge his bets, the chartmaker placed a second magnetic north pole farther south and a bit to the east, where a larger symbol that cartographic historians Helen Wallis and Arthur Robinson describe as "a high rocky island" carries a more confident explanation: "From sure calculations it is here that lies the magnetic pole and the very perfect magnet which draws to itself all others, it being assumed that the prime meridian be where I have placed it." Confronting uncertainty, Mercator used a pair of "extreme positions" to bracket the magnetic pole's true location "until the observations made by seamen have provided more certain information." Too few present-day cartographers, sad to say, are as frank about geographic ambiguity.

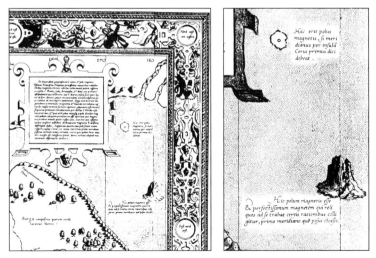

Figure 1.5 Left: The upper-right sheet of Mercator's world map of 1569 depicts two tentative locations for the north magnetic pole. From Krücken, "Ad Usum Navigatum." Right: The symbols in greater detail. From "Gerard Mercator's Map."

Ships carried magnetic compasses as early as the twelfth century, but seamen seldom used them because of an innate mistrust of innovations as well as quirky needles that didn't point directly north. Magnetic declination was not discovered until the fifteenth century, and as Mercator's experience illustrates, geomagnetism proved less well-behaved than sixteenth-century mapmakers had originally believed. Adjustment for geomagnetic distraction was not possible until 1701, when Edmund Halley published a pioneering but simplistic map of isogons (lines of equal magnetic declination) for the Atlantic Ocean (fig. 1.6). Determining a ship's location at sea was equally troublesome. Latitude could be figured simply by sighting on the northern star at night or by measuring the sun's noontime elevation above the horizon, but longitude, calculated from the difference between local

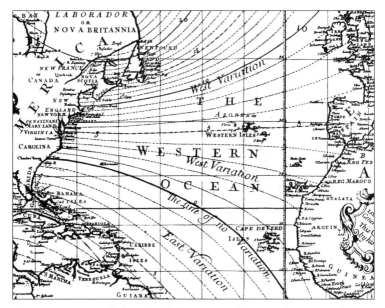

Figure 1.6 A portion of Edmund Halley's 1701 map of magnetic deviations in the Atlantic. Halley plotted his isogons (lines of equal magnetic declination) on a Mercator projection. From Bauer, "Haley's Earliest Equal Variation Chart," facing 33.

time and time at the prime meridian, required a highly accurate chronometer, not available until the mid-eighteenth century, when John Harrison (1693–1776) devised a clock that lost only fifteen seconds in 156 days. The ship's compass, magnetic adjustment, and an accurate chronometer were parts of a puzzle that included Mercator's projection. Not until all the pieces were in place could mariners fully appreciate Mercator charts.

Navigators began to use the Mercator map in the early 1600s, after British geographer Richard Hakluyt (1552–1616) included a world map drawn by Wright in the second edition of his *Principall Navigations, Voiages, Traffiques and Discoveries of the English Nation,* published in 1599. Wright not only corrected inaccuracies in Mercator's grid but provided an updated view of world geography, taken from a 1592 globe by Emery Molyneux (d. 1598/9). Measuring 17 inches (43 cm) tall by 25 inches (64 cm) wide and printed in two sections, the Wright-Molyneux map, as it's often called, is smaller and more readily reproduced, displayed, and archived than Mercator's eighteen-sheet mosaic. According to Robert Karrow, nineteen copies of the Wright-Molyneux map exist, in contrast to only three copies of Mercator's, which is seldom reproduced in one piece because of its size. Despite suggestions that the grid be called the Wright projection, Mercator's name stuck, reinforced no doubt by his impressive contribution as an atlas publisher. Cartographic historians celebrate Gerard Mercator for two epic achievements: his world map of 1569 and his monumental three-volume world atlas, completed in 1595.

Mercator might not have been the first to use the projection that bears his name. In 1511, Erhard Etzlaub (ca. 1460–1532), a Nuremberg compass maker, crafted a portable sundial with a map on its lid. A mere 3.1 inches (80 mm) wide and 4.3 inches (108 mm) tall, Etzlaub's map puts south at the top and covers only Europe and North Africa (fig. 1.7). It lacks a graticule, but latitude gradations at one-degree intervals along the sides and numerical labels every five degrees reflect the progressively spaced parallels of the Mercator grid. This similarity is hardly an accident. Etzlaub produced a similar but slightly larger sundial map two years later and presumably made others that didn't survive. An instru-

Figure 1.7 Latitude gradations along the sides of Erhard Etzlaub's 1511 "compass map" reflect the progressive spacing of parallels characteristic of the Mercator projection. From Kretschmer, "Mercators Bedeutung," 163, fig. 4.

ment maker and physician with an active interest in astronomy and cartography, he produced several other maps, principally woodcuts with south at the top. Especially note-worthy is his 1500 road map of central Europe, cast on a stereographic projection—also conformal—to promote the accurate alignment of compass points with travel directions. According to cartographic historian Brigitte Englisch, his 1511 "compass map" not only was the earliest rectangular conformal projection but also accords exceptionally well with modern versions of the Mercator projection. Englisch argues that Mercator no doubt knew of Etzlaub's invention and that "the projection of varying latitudes should be known as the Etzlaub-Mercator projection."

Wright and Etzlaub are not the only mapmakers in line to share Mercator's fame. Another contender is the unidentified Chinese scholar who drafted the tenth-century Dunhuang star map. According to *The Timetables of Science,* a chronology published in 1988 and cited on several Web sites, the star chart "uses a Mercator projection [and is] the first known use of this kind of map projection." I tracked this assertion no further than the multivolume *History of Cartography,* which includes a black-and-white photo of the narrow, scroll-like map. How the claim arose is a puzzle insofar as the chart contains neither a grid nor marginal tick marks. As a key sentence in its caption tellingly observes: "There is no attempt at a projection on this rather crude chart." Projection guru John Snyder wholly ignored the Dunhuang star chart in his epic history of map projection, in which he noted Etzlaub's "similar projection" but concluded that "the principle remained obscure until Mercator's independent invention."

Anyone who thinks cartographic folklore inflates Mercator's contribution should be mollified if not amused by an offhand comment in the U.S. Coast and Geodetic Survey's bible on map projection, introduced in 1921 and shepherded through numerous revisions by Charles Deetz and Oscar Adams. In discussing the sinusoidal projection, on which converging meridians yield a world map shaped like an antique Christmas ornament, Deetz and Adams noted the occasional use of an alternative name, Sanson-Flamsteed projection, commemorating Nicolas Sanson and John Flamsteed, who used it around 1650 and 1729, respectively. In their opinion, the projection "might well have been termed the 'Mercator equal-area projection' in the first place, from the fact that the early atlases bearing his name gave us the first substantial maps in which it is employed. Mercator's name has, however, been so clearly linked with his nautical conformal projection that it becomes necessary to include with his name the words *equal-area* if we wish to disregard the later claimants of its invention, and call it the *Mercator equal-area projection.*" To underscore the point, they titled the section "Sinusoidal or Mercator Equal-Area Projection."

Whatever its authorship, the better-known Mercator conformal projection gathered adherents among scientists as well as navigators. Noteworthy adoptions include Robert Dudley's pioneering sea atlas of the world, published in 1647, and Edmund Halley's revolutionary maps of the trade winds and magnetic declination, published in 1686 and 1701, respectively. In 1769 the Mercator grid provided the geographic framework for a groundbreaking Gulf Stream chart by Benjamin Franklin and whaling captain Timothy Folger, and in the early nineteenth century it gained wider exposure in the influential line of geography textbooks written and published by Jedidiah Morse, father of portrait painter and telegraphic experimenter Samuel F. B. Morse. In 1919 Vilhelm Bjerknes, the Norwegian meteorologist who discovered fronts and air masses, proposed the Mercator projection as the world standard for weather maps of the tropics, and in 1937 the World Meteorological Committee recognized the importance of conformality on atmospheric maps by endorsing Bjerknes's recommendation. In the commercial sphere, publishers of reference atlases and wall maps adopted the Mercator grid for regional maps of Australia, the Pacific islands, and the world's oceans.

To the disgust of geographic educators, Mercator's grid framed many whole-world maps with no bearing on navigation, weather, or geophysics. As I show in chapter 9, geopolitical motives were apparent in a few cases, but much of the projection's misuse reflects a mix of comfortable familiarity, public ignorance, and institutional inertia. No one was hawking the Mercator brand, at least not overtly, but no one had to—many people who grew up with the map apparently believed this was how a flattened earth should look. How else to explain the ascendancy of an utterly inappropriate perspective and widespread resistance to superior substitutes?

If there is a villain here, it's not Gerard Mercator, who used equal-area maps in his atlases and was quite clear about why he devised a rectangular conformal projection. Wary of wrongheaded finger pointing, Deetz and Adams chided the chartmaker's critics in verse:

> Let none dare to attribute the shame
> Of misuse of projections to Mercator's name;
> But smother quite, and let infamy light
> Upon those who do misuse,
> Publish or recite.

Although educators and scientists understood the problem, few seemed willing to challenge the conventional stupidity.

The most famous attack on the Mercator map's undeserved prominence came well after the tide had turned. In the 1970s German historian Arno Peters (1916–2002) proposed a ludicrously inapt solution now known as the Peters projection. As chapter 11 explains, the Peters map is not only an equal-area map but an exceptionally bad equal-area map that severely distorts the shapes of tropical nations its proponents profess to support. Its popularity among Third World advocacy groups like Oxfam and the World Council of Churches is hard to explain. Perhaps it's a reflection of what I call the Monty Python Effect, named for the parody troupe's well-known transition line, "And now for something completely different." To most people who see it for the first time, the Peters map is indeed different: as figure 1.8 illustrates,

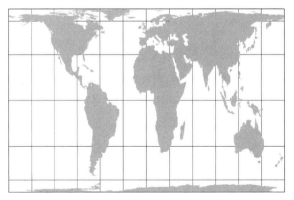

Figure 1.8 Proposed inappropriately as the only suitable substitute for the Mercator map, the Peters projection grossly distorts the shapes of Africa and South America. This example employs a thirty-degree graticule.

Africa and South America look like land masses stretched into sub-mission on a medieval torture rack. In asserting a new solution to an old problem, Peters ignored other, demonstrably better equivalent projections. And in claiming his projection was original, he over-looked an identical map presented in 1855 by James Gall (1808–95), a Scottish clergyman. Dare I say it? Peters had a lot of Gall in as many ways as possible.

Mercator's legacy is much more than the life and works of a Flem-ish chartmaker. As the remaining chapters illustrate, the Mercator projection lies at the intersection of a diverse collection of intriguing tales about navigation, cartographic innovation, military precision, media mischief, and political propaganda.

Early Sailing Charts

As predecessors go, portolan charts are an impressive lot. In addition to having held the mathematically superior Mercator projection at bay for a century or two after its initial presentation in 1569, they attract a far greater following among map historians, who recognize them as a distinct cartographic genre. And as this chapter observes, portolan charts not only taught mariners to rely on sailing charts but also left a legacy of geographic detail for later mapmakers.

It's easy to treat portolan charts as both enigma and innovation. They appeared suddenly in the late thirteenth century with crisscrossed rhumb lines and abundant place names, all in sharp contrast to the prevailing religious cartography typified by small, sparse, east-up world maps centered on Jerusalem. Unlike the medieval *mappaemundi,* which were largely inspirational, portolan charts were practical tools for crossing open waters. And unlike the well-documented publication of Gerard Mercator's world map, the murky origin of the portolan charts has invited much speculation, not likely to be

resolved, about whether Italians or Catalan Spaniards crafted the ultimate prototype, which historians have yet to find.

In their handbook of cartographic innovations, map historians
Helen Wallis and Arthur Robinson list four key characteristics of portolan charts. Foremost is the web of intersecting rhumb lines, typically
originating on the circumference of a circle, around which sixteen
equally spaced points represent the eight principal wind directions
(N, NE, E, SE, S, SW, W, and NW) and the eight half-winds (NNE, ENE,
ESE, . . .) of the mariner's compass (fig. 2.1). On most charts the circle
is readily apparent in the points at which rhumb lines converge like
spokes in a wheel. Look closely at the portolan chart in figure 2.2,
which covers the western Mediterranean, and you'll see traces of a
large circle centered at the middle of the chart and touching the top
and bottom edges. Rhumb lines also converge at the circle's center,
and at the lower right, over North Africa, one of the sixteen intersection points on its perimeter serves as a compass rose. On some oblong
portolan charts, like the example in figure 2.3, adjacent circles cover
eastern and western parts of the map.

Closer inspection of the chart in figure 2.2 reveals a second distin-
guishing trait: an abundance of closely spaced, hand-lettered place
names perpendicular to the shoreline and always inland, to avoid conflict with coastal details. Additional labels over water identify small is-

Figure 2.1 Subdivision of the sixteen
points of this portolan compass rose
added the sixteen quarter-winds represented by plain rhumb lines. From
Stevenson, *Portolan Charts*, viii.

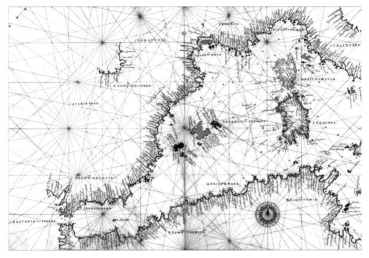

Figure 2.2 A much-reduced view of coastal features, rhumb lines, and place names on a Mediterranean chart in a 1544 portolan atlas by Battista Agnese. From Agnese, *Portolan Atlas,* image 9.

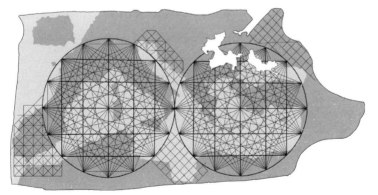

Figure 2.3 Shorelines, rhumb lines, circles, and peripheral grids (believed to indicate scale) from the *Carte Pisane* (ca. 1300). Traced from Bagrow, *History of Cartography,* pl. 32, with guidance from a line drawing in Lanman, *On the Origin of Portolan Charts,* 35.

lands. Because chartmakers inked in these names one after the other in a continuous coastwise sequence, labels appear inverted where the shoreline reverses direction. A third characteristic is color-coded names and directions. More important places, labeled in red, stand out from less significant neighbors, lettered in black. Color also reduces confusion among rhumb lines, inked in black or brown for the eight principal winds, in green for the eight half-winds, and in red for the sixteen interspersed quarter-winds. The fourth trait is a functional generalization that rounds minor coastal irregularities, overstates bays and headlands, and uses crosses and dots to point out rocks and shoals. Except for lavishly decorated versions intended for royal collectors, portolan charts showed what mariners needed to know and not much else.

Inked on treated animal skin called vellum, portolan charts withstood rough handling at sea better than paper navigation charts, which did not become common until the eighteenth century. Animal hides were especially suited to the Mediterranean's pronounced east–west elongation. After splitting the calf's or sheep's skin along the stomach, the vellum maker removed the appendages and head but kept the neck, which formed the noticeably narrowed end of a large oblong drawing surface. The typical portolan chart is drawn on a single skin with the tapered end pointing west, to accommodate the Mediterranean's narrowed reach toward the Atlantic. The flesh side of the skin provided a smooth writing surface; younger animals, with fewer scars, were preferred. Treatment included soaking the hide in lime, scraping off hair and flesh, stretching over a drying frame, rubbing with pumice to smooth the surface, and massaging with chalk to create a neutral, off-white background. Although the charts could be rolled for easy storage—like a thin leather glove, vellum is flexible— some were mounted on wood or cardboard to prevent shrinkage.

Medieval chartmakers are not wholly anonymous. Tony Campbell, the British Library's former map librarian who wrote the chapter on portolan charts for the multivolume *History of Cartography,* lists forty-six individuals known to have produced portolan maps or atlases before 1500. Especially noteworthy are Pietro Vesconte, a Genoese mapmaker whose earliest known nautical map is a 1311 chart of the Mediterranean and the Black Sea, and Giovanni da Carignano, a Genoese abbot once credited with the earliest dated portolan chart, believed to have been drafted around 1300. No one questions Carignano's authorship of the chart, which was destroyed during World War II, but comparison of photographic copies with other maps of the period reveals places names not widely known or used until the 1320s. Cartography was not Carignano's vocation, but by the late fourteenth century demand for sailing charts was supporting specialist chartmakers in the Italian ports of Genoa and Venice as well as their Catalan counterparts of Barcelona and Majorca.

At least a few medieval chartmakers benefited from an edict endorsing navigation maps. In 1354 King Peter of Aragon ordered all ships to carry two portolan charts, the second perhaps as backup if the other were ruined. Peter's ordinance reflected the charts' value as navigation aids as well as the consequences of a ship foundering or getting lost. The earliest surviving record of a chart used at sea is an account of a 1270 voyage by France's King Louis IX. Because of rough weather the captain decided to seek shelter at Cagliari, in Sardinia, and brought out a chart to reassure the frightened monarch that land was nearby.

The oldest known portolan chart is the *Carte Pisane,* drafted around 1290 in Genoa but named after Pisa, where it was discovered. Shown schematically in figure 2.3, the chart measures 20 by 41 inches (50 by 104 cm), encompasses the Mediterranean and part of the Black Sea, and includes all four characteristics of its genre. Separate circles anchor two networks of rhumb lines. Hidden on later charts, the circles here are inked in and obvious. Beyond the circles are several squarish grids, with no apparent role. Although seventeenth-century mapmakers used temporary grids, sketched in pencil, as guides for copying features from other charts, erasable pencils were not available until the sixteenth century. Tattered edges and missing fragments of vellum toward the upper right reflect repeated handling. Acquired in 1839 by the Bibliothèque Nationale, the *Carte Pisane* is a lucky survivor. Campbell, who uncovered fewer than two hundred pre-1500 portolan charts in public and private collections, dedicated his chapter to "the thousands of ordinary charts that served their purpose and then perished."

Although scholars have yet to uncover a detailed description of medieval chartmaking, they're certain that portolan charts were copied by hand from existing charts. Microscopic analysis of inked lines and tiny pinholes indicates that chartmakers first laid out the rhumb circle by using dividers (an instrument with two sharp points for transferring exact dimensions) to mark its center and sixteen equally spaced points on its circumference. Using the pinpricks as guides, artisans inked in the network of rhumb lines with pen and straightedge. They then transferred the shorelines from a master map, but exactly how remains a mystery. Some chartmakers apparently forced a fine powder through small holes in a master pattern placed over the fresh vellum, some used a crude form of carbon paper, and some are alleged to have anchored the master map on a transparent frame or table, placed the vellum on top, positioned a strong light source on the opposite side, and traced coastlines and other features directly. Still others might have been exceptionally good at visual transfer—what my cartography students call "eyeballing it." Once the shorelines were laid down, transferring the place names was a straightforward yet painstaking process.

The prevalence of copying raises questions about the ultimate master chart: who crafted it, when, and how? Although map historians hold little hope of identifying the first chartmaker, they're certain the prototype portolan chart—if indeed there was only one—was compiled from maps of smaller areas based on books of sailing directions called *portolani.* Written to help seamen find ports and avoid hazards along the Mediterranean coast, these medieval Italian sailing guides have an equally obscure origin. Although sailors had been taking notes on coastal navigation for over a millennium, pilot books with distances and bearings as well as shoreline narratives emerged at about the same time as the portolan charts. Or perhaps a bit before: extant portolan charts greatly outnumber surviving *portolani,* which were not decorated and never caught the fancy of royal collectors.

Wary of untested assumptions, Jonathan Lanman, a retired medical researcher and map collector, compiled sailing maps from the *Lo Compasso de Navigare*, a pilot book from the late thirteenth century, and the *Parma-Magliabecchi Portolano*, from the fifteenth century. Although fragments of older sailing guides exist, these were the earliest, most complete examples he could locate. To assess the cartographic validity of their sailing directions, he reconstructed the Mediterranean shoreline by chaining together straight-line segments based on distances reported in Italian sea miles, 1 sea mile equaling 0.67 nautical miles (1.23 km), and bearings based on a thirty-two-point compass rose. Rotation of the resulting plots and careful alignment with the present-day shoreline revealed a realistic representation of the Mediterranean coast. Despite less than perfect matches, Dr. Lanman demonstrated that the information in the sailing guides was fully adequate for drawing dependable portolan charts.

Curious about the roles of map projection and magnetic declination, Lanman examined the geometric accuracy of the *Carte Pisane* and a second chart drawn in 1559 by Matteo Prunes, a Majorcan chartmaker. Although cartographic historians generally consider portolan charts "projectionless" for lack of a graticule, Lanman suggested they were "drawn on a square grid" noticeably skewed as a result of magnetic declination. Although evidence of an overt grid is speculative—Lanman's argument rests largely on small squares within the rhumb circles of the *Carte Pisane* and few other charts—locally reliable shapes reflect at least an unconscious appreciation of conformality, a key property of the Mercator projection. Researchers who have confirmed this proto-conformality (my term) include Waldo Tobler, a pioneer in computer cartography, who observed a strong similarity between a 1468 chart by Majorcan chartmaker Petrus Roselli and an oblique Mercator projection. And in a cartometric analysis of twenty-six charts, Scott Loomer, a cartography instructor at West Point, found strong correlations with conformality and straight loxodromes—exactly the properties needed for reliable navigation over open waters. Because a medieval sailing chart typically covered a small area, its informal, ad hoc projection was not a serious weakness.

Some historians recognize the 4 by 4 grids within the *Carte Pi-sane*'s rhumb circles as linear scales, running vertically as well as horizontally. Intersecting grid lines divide a distance of roughly 200 miles into four equal parts, and horizontal and vertical scales that are similar—or would be if the interior elements were perfect squares—signify the chartmaker's unconscious pursuit of conformality. At least that's how map historians interpret the grid. Unlike the scale bars on contemporary maps, scales on portolan charts didn't specify distance.

The role of the magnetic compass in the compilation and use of portolan charts remains contentious. Did the compass play an important part in the compilation of early prototypes, or did it merely contribute to a more effective use of sailing charts and later updates? According to Lanman, the orientation of map features accords well with historic trends in magnetic declination. But in Tony Campbell's view the jury is still out. The magnetic compass was in use by the thirteenth century, but it's questionable whether instruments available around 1290 were sufficiently reliable to have contributed significantly to either the *Carte Pisane* or contemporary *portolani.* Magnetic variation, which could provide a clue, is difficult to reconstruct, especially before 1600. Although a westward increase in magnetic deviation in the region was apparent until the seventeenth century, local magnetic anomalies thwart a reliable reconstruction of local details. What's certain is that chartmakers corrected their bearings after better measurements became available around 1600.

Four centuries of portolan charts document European exploration of the African, American, and Asian coasts as well as advances in seamanship in England, Portugal, and what is now the Netherlands. For an appreciation of these improvements, compare the vague rendering of the Mediterranean coast on the *Carte Pisane* (see fig. 2.3) with the more detailed shorelines in the 1544 map by Venetian mapmaker Battista Agnese (see fig. 2.2). The more recent map is a double-page spread from a portolan atlas in the U.S. Library of Congress's cartographic collection. Several of the atlas's nine charts encompass the east and west coasts of North and South America, and a world map depicts the global journey of Ferdinand Magellan's crew—the explorer died en route—

as well as a meandering course from Spain to Panama and then down the coast to Peru. As the charted world expanded beyond the Mediterranean, navigators found the atlas format, with maps on vellum bound in leather, convenient for protecting their charts as well as accommodating new knowledge too detailed for a single map.

Expansion of detailed coverage into the Atlantic encouraged cartographers to correct scale disparities between the charts' Atlantic and Mediterranean sections. Because pilot books for these areas had been compiled independently, with no attempt to resolve inconsistencies, early portolan charts underestimated distances along the North Atlantic coast by 16 to 30 percent relative to distances in the Mediterranean. These discrepancies persisted until 1403, when Francesco Beccari responded to feedback from mariners with a new chart that also corrected another reported deficiency. As the Genoese chartmaker's inscription reveals, "It was several times reported to me by many owners, skippers and sailors proficient in the navigational art that the island of Sardinia ... was not placed on the charts in its proper place. Having listened to the aforesaid persons I placed the said island in the present chart in the proper place." Several decades passed before other chartmakers adopted Beccari's adjustments.

Since portolan charts were constructed from bearings and distances rather than a determination of geographic coordinates, they lacked indications of latitude and longitude and explicit projections. In the sixteenth century, latitude scales made a halting appearance on sailing charts, but even then, they were simply laid over the framework of rhumb lines, rather than integrated with it. Figure 2.4, a chart from a 1582 atlas by Spanish cartographer Giovanni Martines, shows this disconnect. The north–south and east–west lines on the chart do not represent particular meridians or latitudes; they are simply the extensions of these cardinal directions from the various wind roses. Even so, a navigator with dividers could determine his destination's latitude, use a quadrant or astrolabe (instruments for measuring latitude at sea) to guide him north or south to the right parallel, and then sail due east or west to the intended port. Mariners call this parallel sailing.

Figure 2.4 Portolan chart from a 1582 atlas by the Spanish cartographer Giovanni Martines. From Stevenson, *Portolan Charts*, pl. 18.

Medieval chartmakers who included a graticule were usually more precise. The result was an equirectangular projection, characterized by evenly spaced parallels intersecting evenly spaced meridians at right angles. On an equirectangular rendering of the globe, north–south scale is constant and correct throughout, but because the meridians do not converge, east–west scale is generally distorted, particularly at the poles, where the projection stretches a point into a line as long as the equator.

As figure 2.5 illustrates, an equirectangular framework maps the sphere onto a cylinder sharing the same axis. Sphere and cylinder touch along one or two *standard parallels,* which are lines of true east–west scale. In the *tangent* case, with the sphere merely touching the cylinder, the equator is the map's sole standard parallel, and distortion is lowest in its vicinity. In the *secant* case, the cylinder penetrates the sphere along two standard parallels, with equal but opposite latitudes, and distortion is lowest in their vicinity. This effect is illustrated in the right part of figure 2.5, where an equirectangular projection "secant at 45°" has standard parallels at 45° N and 45° S. The map's grid cells are compressed noticeably in the east–west direction because a degree of longitude at 45° is only 70.7 percent as long on the

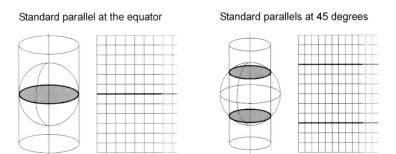

Standard parallel at the equator   Standard parallels at 45 degrees

Figure 2.5 The tangent case (left) of the equirectangular projection places a single standard parallel at the equator and yields a square grid, whereas the secant case (right) results in two standard parallels (each forty-five degrees from the equator in this case) and a nonsquare grid. In this example meridians and parallels are fifteen degrees apart.

globe as a degree of latitude. Both cases, tangent and secant, have true north–south scale throughout, and their meridians are identical in length.

Because distortion increases with distance from a standard line—as I show later, meridians and other lines can also be "standard"—an equirectangular projection with a standard parallel at 39° N, the approximate latitude of Majorca, provides a generally low-distortion portrait of the Mediterranean, no part of which lies more than nine degrees away from this line of true east–west scale. Although the map distorts distances and bearings, the deformation is less troublesome than the imprecise positions of the features on medieval sailing charts.

Devised around AD 100 by Marinus of Tyre, a predecessor of the Egyptian astronomer-geographer Claudius Ptolemy, the equirectangular projection is the oldest and most straightforward cylindrical projection. Although Ptolemy deemed it suitable only for regional maps, equirectangular grids simplified the plotting of points identified by latitude and longitude. In the late fifteenth century Portuguese chartmakers began to use a square-grid variant called the plate carrée or plane chart—essentially the tangent case of the equirectangular projection. With true scale along the equator and all meridians, the plane chart (fig. 2.6) was less suitable for navigation maps of the Mediter-

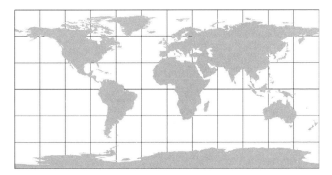

Figure 2.6 A plate carrée, or plane chart, (shown here with a thirty-degree graticule) is the tangent case of the equirectangular projection.

ranean than an equirectangular projection secant within the region, but by 1500 explorers were venturing across and beyond the Atlantic Ocean and their royal and mercantile sponsors needed maps global in scope. As this latter audience grew in size and importance, the plane chart provided a convenient framework for cartographic milestones like Martin Waldseemüller's 1516 *Carta Marina* and Diego Ribero's 1529 *Carta Universal.* Although ill-suited for plotting rhumb lines and estimating bearings, the plane chart was nonetheless useful for parallel sailing and sketching coastlines. Because tradition-bound mariners learned to live with its distortions, the plane chart dominated nautical charting for over a century after Mercator introduced his demonstrably superior 1569 world map.

3

Mercator's Résumé

Gerard Mercator was more than just a mapmaker. Although biographical dictionaries accustomed to single occupations typically treat him as merely a cartographer or a geographer, Mercator distinguished himself at various times as a calligrapher, an engraver, a maker of scientific instruments, and a publisher. No less impressive are his deep interests in mathematics, astronomy, cosmography, terrestrial magnetism, history, philosophy, and theology. Although biographers lament the lack of diaries, account books, and carefully archived personal correspondence, the historical record reveals Mercator as an introspective and energetic chap who was competent in science, honest and well liked, technically savvy and clever with his hands, curious about the world around him, successful as an entrepreneur, and well positioned to make a pair of substantial contributions to mapmaking.

Mercator's first biographer was Walter Ghim, his neighbor in Duisburg, the small German city where he lived from 1552 until his death in 1594. A twelve-term mayor of the town, Ghim contributed a short biography to the 1595 edition of Mercator's *Atlas*, published posthumously by his youngest son, Rumold. Ghim's essay is more a long obituary than a critical biography. The mayor praises Mercator as a "remarkable and distinguished man," notes his "mild character and honest way of life," and provides dates and other details for key events in the cartographer's career. Thus we learn that Gerard Mercator was born at approximately 6 a.m. on March 5, 1512, in Rupelmonde, Flanders, where his parents Hubert and Emerentiana were visiting Hubert's brother, Gisbert Mercator, "the energetic priest of that city." (Flanders is roughly coincident with the northern part of present-day Belgium, and as figure 3.1 shows, the village of Rupelmonde is about ten miles southwest of Antwerp.) He died "82 years, 37 weeks, and 6 hours" later—a remarkably long life for the sixteenth century—after coping in his final years with partial paralysis and a cerebral hemorrhage. Ghim offers a detailed description of Mercator's failing health and last rites but says little about the mapmaker's early life.

Scholarly interpretations of sixteenth-century Flanders helped historian of calligraphy Arthur Osley paint a richer picture. Although Mercator's parents had little money—his father was a shoemaker and small farmer—Gisbert was at least better connected. Through his uncle's influence, Gerard was enrolled at age fifteen in the distinguished monastic school at 's-Hertogenbosch run by the Brethren of the Common Life, who accepted poor but bright boys willing to train for the priesthood. The brothers specialized in copying sacred texts, and their school excelled at teaching penmanship. In addition to learning Christian theology and Latin, Mercator developed a practical and lasting interest in the elegant italic script in which he engraved place names and interpretative text for his maps. He considered italic lettering more appropriate for scholarly writing than Gothic and other less formal (and often less legible) styles of handwriting, and in 1540 he published *Literarum latinarum, quas Italicas cursoriasque vocant, scribendarum ratio* (How to Write the Latin Letters Which They Call Italic or Cursive), a short manual that was influential in the adoption of italic lettering in cartography.

Various renderings of Mercator's name invite confusion. Although his German father apparently went by Hubert Cremer, vernacular

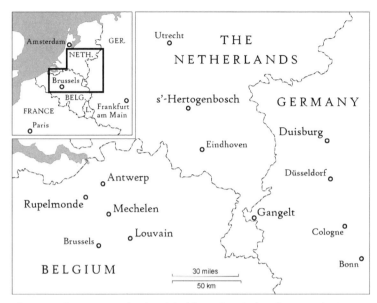

Figure 3.1 Places Mercator lived or visited (larger lettering), with present-day international boundaries and additional cities (smaller labels) as a frame of reference.

versions of the family name include de Cremer, Kramer, and Kremer. *Krämer* (the modern spelling) is the German word for merchant or shopkeeper, *Cremer* is its Dutch equivalent, and *Mercator* is the Latin version, which the future mapmaker adopted at 's-Hertogenbosch. (Latin was the language of Europe's educated elite, and young scholars routinely latinized their names.) Although Gerhard Cremer and Gerardus (or Gerhardus) Mercator might be more historically correct, American and British cartographic historians prefer the partly anglicized Gerard Mercator. A reasonable compromise, I'm sure, as an obsessive purist would need to write awkwardly about Gerardus Mercator Rupelmundanus (Gerard Mercator of Rupelmonde), the name under which Mercator enrolled at the University of Louvain in 1530 and published his epic world atlas.

At Louvain Mercator studied humanities and philosophy, attended lectures by the brilliant mathematician and astronomer Gemma Fri-

sius (1508–55), and received a master's degree in 1532. With his religious faith challenged by contradictions between biblical accounts of
creation and Aristotle's writings, Mercator occasionally felt stifled at
Louvain, where doubt was akin to heresy. He began corresponding
with a group of Franciscan preachers living in Antwerp and Mechelen
(see fig. 3.1), and visited them several times to discuss theology and
science. His confidants included Franciscus Monachus (ca. 1490–
1565), a prominent geographer who produced a terrestrial globe
around 1520 and is a plausible source of Mercator's knowledge of
northern lands. Although his absences from Louvain aroused suspi-
cion, Mercator eventually resolved his concerns over the conflicting
interpretations and, according to Osley, "emerged with strong Christian convictions, which remained with him."

Reluctant to leave Louvain, Mercator pursued an academic ap-
prenticeship centuries before the modern university gave us post-
graduate education. In addition to convincing Frisius to instruct him
in astronomy and geography, Mercator and his tutor persuaded Gas-
par van der Heyden, a local goldsmith and engraver, to let Mercator
use his workshop for making globes and scientific instruments. The
three apparently collaborated on numerous projects, including maps
and surgical instruments—Frisius was also a physician—and the future mapmaker either contributed to or witnessed all phases, from design to marketing. As Osley observes, by age twenty-four Mercator
had become "a superb engraver, an outstanding calligrapher, and one
of the leading scientific instrument makers of his time." And as his
later works attest, skill in engraving gradations and labels on brass
and copper instruments proved useful in making printing plates for
maps and globe gores.

An energetic learner, Mercator progressed quickly from globes to
flat maps and from engraving to full authorship. In 1536 he engraved
the italic lettering for Frisius's terrestrial globe, which was assembled
by pasting twelve printed gores onto a spherical papier-mâché shell
nearly 15 inches (37 cm) in diameter. His role expanded from engraver to coauthor with the publication a year later of Frisius's celestial globe, similar in size and manufacture. In 1537 he also authored

and published his own map, a 17 by 39 inch (43 by 98 cm) carto-graphic portrait of Palestine engraved on copper and printed as six sheets, which formed a wall-size map when glued together. Mercator's enduring interest in religion was no doubt a key motivation. Although he cites Jacob Zeigler as his principal source, the small map included with Zeigler's book on the Holy Land, published five years earlier, is comparatively sketchy. Cartographic historian Robert Karrow, who la-beled the map a "commercial success," notes that it remained in print for at least four decades and provided the geographic details for Pales-tine for Mercator's epic world map of 1569.

In 1538 Mercator published a 14 by 21 inch (36 by 55 cm) world map, laid down on the double cordiform (double heart-shaped) pro-jection (fig. 3.2) pioneered in 1531 by the French mathematician Oronce Fine (1494–1555). Although Mercator borrowed the geo-graphic framework from Fine, his map is more similar in content to Frisius's terrestrial globe. As close examination of its features and place names reveals, he consulted additional sources but was the first

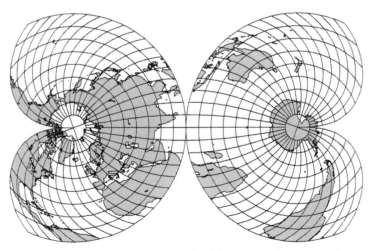

Figure 3.2 A modern rendering by John Snyder of the double-cordiform projection used for Mercator's 1538 world map. Reduced slightly from Snyder, *Flattening the Earth,* 37, fig. 1.27.

to identify North and South America as separate continents. Also noteworthy are the suggestion of a Northwest Passage and the separation of Asia and North America, typically attached on early-sixteenth-century world maps. Aware of the uncertainty of some delineations, he scrupulously differentiated known, previously mapped coastlines from their more speculative counterparts in areas largely unexplored.

Mercator's next publication was a detailed 34 by 46 inch (87 by 117 cm) map of Flanders, printed as four sheets in 1540. Prepared at the urging of Flemish merchants, the map was based on precise trigonometric and field surveys. Although some historians attribute the measurements to Mercator, who no doubt engraved the copper plates, others question whether the impoverished artisan had the time and resources for extensive fieldwork during the harsh winters of 1537–38 and 1539–40. A key skeptic is Rolf Kirmse, who observed that the distances portrayed are off by only 3.4 percent on average and that the average error of the angles is a mere 2° 20′. According to Kirmse, the timing of the surveys and their high level of accuracy point to Jacob van Deventer (ca. 1500–1575), a Dutch mapmaker who lived in Mechelen in the late 1530s and later produced a unique collection of town plans of the Netherlands for the king of Spain. Whoever the surveyor, there is no dispute about the map's success and influence. Among the fifteen subsequent editions published between 1555 and 1594 is a smaller adaptation included in the 1570 world atlas by Abraham Ortelius (1527–98), a genial contemporary of Mercator.

In August 1536 Mercator married Barbara Schellekens, and the following year Barbara gave birth to their first son, Arnold. The couple eventually had six children, three boys and three girls. All three sons became mapmakers for a time at least, and Rumold (ca. 1541–1600), their youngest, became his father's representative in England and supervised publication of the first complete edition of the Mercator world atlas.

Although prosperous and comparatively erudite, sixteenth-century Flanders was frequently engulfed in conflict between Protestant reformers and Catholic traditionalists, who in 1544 began a brutal effort to suppress Protestantism. Mercator's letters to the friars in Mechelen as well as his more recent travels attracted the attention of religious extremists, who imprisoned him at Rupelmonde in March 1544. The zealots also held forty-two other suspects, including Joannes Drosius, to whom Mercator had dedicated his 1538 world map. Although protests by the mapmaker's friends, colleagues, town officials, and a local priest won his release seven months later for lack of evidence, four of his fellow detainees were beheaded, burned at the stake, or buried alive.

Mercator's religion remains ambiguous. Some writers consider him a Protestant (possibly a Lutheran convert), while others insist he remained a committed Catholic. Ghim and Osley ignore the mapmaker's church affiliation altogether, Karrow confesses uncertainty, and the late Richard Westfall, who compiled the entry on Mercator for the Catalog of the Scientific Community Web site, emphatically states, "I find it impossible to tell." Mercator was released from his imprisonment into Catholic territory, Westfall notes, but eight years later he left Louvain for Duisburg, in Cleve (a German duchy about fifty miles east of Flanders), which was Protestant. Even so, Catholic patrons continued to sponsor his projects and buy his maps.

Although religious unrest or outright persecution might have precipitated the move, the immediate incentive was a job offer from William, Duke of Cleve, who planned to open a university in Duisburg. Although the duke's academy never developed, royal and commercial patrons continued to underwrite Mercator's globes, maps, and scientific instruments. Especially significant is his 1554 map of Europe, which he started in Louvain. Engraved in copper and printed as fifteen separate sheets, the entire map measures 47 by 58 inches (120 by 147 cm) and, according to the ever enthusiastic Walter Ghim, a revised edition published in 1572 "attracted more praise from scholars everywhere than any similar geographical work which has ever been brought out."

The 1554 edition's portrayal of Britain underscores the difficulty of obtaining accurate geographic information about a country that feared invasion. According to Peter Barber, the British Library's expert in medieval and early modern maps, Mercator relied heavily on existing maps, including a 1546 map of England published in Rome by George Lily, as well as reports from various unnamed correspondents, including the British astronomer-mathematician John Dee, who lived in Louvain from 1538 to 1540. Although his correspondents helped him add place names and refine coastlines, Mercator's treatment does not mirror the markedly more accurate geometry of unpublished British surveys of the late 1540s and early 1550s. More surprising is the omission of several bishoprics that Henry VIII had established after he broke with Rome—surprising because Mercator, now living in Duisburg, had little to fear from church authorities. In Barber's view, the omission reflects either ignorance of the bishoprics or a reluctance to antagonize a generous supporter, Cardinal Granvelle, to whom Mercator dedicated the map.

More impressively accurate is Mercator's 1564 map of England, Scotland, and Ireland, printed on eight sheets, which compose a 35 by 50 inch (88 by 127 cm) wall map. A curious inscription attributes its content to a prototype mysteriously acquired from an anonymous acquaintance. According to Ghim, "a distinguished friend sent Mercator from England a map of the British Isles, which he had compiled with immense industry and the utmost accuracy, with a request that he should engrave it." Neither Mercator nor Ghim named the source, whose identity sparked the curiosity of map historians who, as Barber tells it, eagerly enlisted in a game of "find the friend." After analyzing place names, shapes, and other details together and carefully assessing information available to plausible informants, Barber attributed the draft to John Elder, a Scottish Catholic who traveled freely between England and mainland Europe. Elder had access to the Royal Library, where he apparently compiled the map from ostensibly top-secret drawings by English surveyors. According to Barber's hypothesis, Elder left England in late 1561, amid growing hostility between the Catholic and Protestant supporters of Mary Stuart and Elizabeth I, and gave the map to Cardinal de Lorraine, who persuaded Mercator to make the engraving.

Although powerful patrons like the Cardinal no doubt initiated specific projects, serendipitous influences were at least equally important. For example, Mercator's famous 1569 world map, discussed in greater detail in the next chapter, was at least partly encouraged by his appointment to teach mathematics, as a part-time volunteer, in the gymnasium (high school) established by Duisburg's city council in 1559. Mercator designed a three-year course that included geometry, surveying, and mathematical astronomy, and he taught the entire sequence once before surrendering the position to his second son, Bartholomew. A second example is his appointment around 1564 as cosmographer to the Duke of Jülich, Cleve, and Berg. According to Karrow, this nomination inspired Mercator to plan an enormous series of works on geography, cosmography, and history. The first part to be published was the *Chronology* (1569), an attempt to establish an accurate framework for world history. The *Chronology* included tables of solar and lunar eclipses and a conscientiously researched chronological list of political, cultural, scientific, and biblical events. Committed to completeness, Mercator earned a place on the Church's list of banned books by including events associated with Martin Luther and a few other heretics.

As a second installment of his vast, comprehensive work, Mercator published an authentic version of Ptolemy's *Geography,* deliberately devoid of the distracting interpretations and misinterpretations by earlier editors intent on improving the Egyptian geographer's seminal work. Mercator's goal was an accurate portrait of Ptolemy's second-century view of the world. To understand the present, the mapmaker believed, one must appreciate the past. The atlas, published in 1578, included Ptolemy's twenty-seven maps, carefully restored, handsomely engraved, and supplemented by an index of place names and an enlarged boundary map of the Nile Delta. The maps vary slightly in size, with the typical display measuring approximately 13 by 18 inches (34 by 46 cm). Seven subsequent editions, published between 1584 and 1730, attest to the book's importance to scholars. An engraved portrait of Mercator holding a globe and dividers (fig. 3.3) suggests that the mapmaker, now in his seventies, had become a brand name in geographic publishing.

While working on Ptolemy's *Geography,* Mercator had started to

Figure 3.3 Based on a 1574 portrait, this elegant engraving of Gerard Mercator measuring a globe was first printed in the 1584 edition of Ptolemy's *Geography*. It also appeared in the 1595 edition of Mercator's *Atlas*. From Averdunk and Müller-Reinhard, "Gerard Mercator," frontispiece.

compile maps for his celebrated world atlas, which would provide the modern geographical component of the massive treatise he envisioned. Resolving discrepancies between sources and engraving most of the plates himself was a slow process, especially for a seventy-year-old mapmaker. Trading off delay and fragmentation, he published *Atlas sive Cosmographiæ Meditationes de Fabrica Mundi et Fabricati Figura* (Atlas, or Cosmographic Meditations on the Fabric of the World and the Figure of the Fabrick'd) in three installments: a 1585 edition, with 51 maps focused largely on France, Germany, and the Low Countries; a 1589 volume, with 23 maps taking in Italy and Greece; and the complete, 1595 edition, which reprinted the 74 maps issued earlier and added 28 new maps covering most of the remaining parts of Europe.

Because the atlas lacks detailed maps of Spain and Portugal, "complete" is misleading. Mercator no doubt desired a more comprehensive treatment of Europe, but time was running out. Weakened by strokes in 1590 and 1593, he died on December 2, 1594, leaving completion to his son Rumold and grandsons Gerard, Johann, and Michael. In addition to supervising printing, Rumold authored a world map and a regional map of Europe, Gerard signed regional maps of Africa and Asia, and Michael contributed a map of America. The project also provided employment for local artisans, who hand-colored the maps. Like other mapmakers, Mercator relied on colorists, mostly women, to enhance his otherwise bland line engravings.

What took so long? The late Clara LeGear, an atlas authority at the U.S. Library of Congress, identified four impediments: Mercator's need to support himself with other projects, the difficulty of obtaining reliable geographic details, the slow pace of meticulous map engraving, and a shortage of skilled copperplate engravers. Mercator not only compiled all the maps for the atlas but also engraved the printing plates, with only occasional help from his grandson Johann and Frans Hogenberg, a skilled artisan who engraved most of the seventy maps for *Theatrum Orbis Terrarum* (Theater of the Whole World), published in 1570 by Abraham Ortelius, a publisher and map seller living in Antwerp.

Although a competitor, Ortelius was also a close friend of Mercator. So close, according to Walter Ghim, that Mercator deliberately delayed his own atlas. As Ghim tells it, Mercator "had drawn up a considerable number of models with his pen" and could easily have had them engraved. Yet he held up publication until Ortelius "had sold a large quantity of *Theatrum* . . . and had subsequently increased his fortune with the profits from it." A nice story, perhaps, but the tedium of map engraving as well as the fifteen years between *Theatrum* and the first installment of Mercator's *Atlas* suggests Ghim was spinning a yarn.

In pioneering the notion of a consciously organized book of mainly maps with a standard format printed in uniform editions of several hundred copies, Ortelius has a stronger claim than Mercator to the title Father of the Modern World Atlas. According to map historian Jim Akerman, the innovative ingredient was *Theatrum*'s structure, not its format. After all, bound collections of portolan charts copied by hand had been around for more than a century, and books of printed maps published by Martin Waldseemüller (1470–1522) and others in the early sixteenth century clearly qualify as atlases. What is noteworthy is Ortelius's demonstration of atlas making as a systematic process orchestrated by an editor who selects information, standardizes content, and maintains quality.

Ortelius and Mercator had decidedly different views of the editor's role. Whereas Ortelius relied largely on readily available sources, which he selected for reengraving, Mercator energetically sought new source materials and authored original maps, which he personally designed and engraved. Unencumbered by this spirit of scholarship, *Theatrum* not only beat *Atlas* onto the market but was so much more successful at the outset that Akerman considers it "remarkable that Mercator's name [for a book of maps] should have eventually triumphed."

Remarkable perhaps, but hardly inexplicable. The "Atlas" of Mercator's title commemorates an ancient ruler of Mauritania. In classical mythology, the immortal Atlas was forced to atone for his role in an unsuccessful revolt by supporting the heavens on his shoulders. In

Mercator's interpretation, Atlas was really a mere mortal magnified to legendary proportions for his accomplishments in science and philosophy. Although Mercator's mythology is questionable, Atlas as a geographer and cosmographer provided an appropriate visual metaphor for the title page (fig. 3.4) of a massive work based on the hard work and persistence of the first truly hands-on atlas editor.

As a word for a book of maps, *atlas* might have vanished shortly after Mercator's grandsons brought out a second complete edition of

Figure 3.4 The title page of Mercator's 1595 *Atlas* honors the mythic Atlas.

GERARDUS MERCATOR NATUS·   IUDOCUS HONDIUS NATUS IN
RUPELMUNDÆ.III NON.MARTII ANNO   PAGO FLANDRIÆ. DICTO WACKENE XVI
CIƆIƆXII:VIXIT ANN.LXXXII.M.VIII.D.   KALEND.NOVEMBRIS ANNO CIƆIƆLXIII:
XXVI:DENATUS IV NON.DECEMBRIS   VIXIT ANN.XLVII.M.VII.D.XXIX:DENAT?
ANNO CIƆIƆXCIV.   US XIV KAL.MARTII ANNO CIƆIƆCXII.

Figure 3.5 In the expanded edition of Mercator's *Atlas* published in 1606 by Jodocus
Hondius and his sons, this engraved portrait of Mercator and Hondius signified the
merger of two important cartographic trademarks. From Averdunk and Müller-
Reinhard, "Gerard Mercator," pl. 18.

the *Atlas* in 1602. Apparently disappointed by sales, they sold the
plates to the family of Jodocus Hondius (1563–1612), who ran a successful engraving and publishing business in Amsterdam. Hondius
and his sons had a two-fold strategy for challenging the less meaty but
still popular *Theatrum.* In 1606 they published a new, more geographically complete edition with forty additional maps. Recognizing the
value of a brand name, Hondius listed Mercator as the author and
himself as the publisher. A contrived engraving of the two collabora-
tors seated at a table with globes and dividers (fig. 3.5) reinforced the
continuity. To lower the cost of engraving, printing, and hand coloring, the *Atlas* maps, which measured about 14 by 18 inches, were simplified and reengraved to roughly 7 by 9 inches and published as the
Atlas minor, a less expensive version introduced in 1607 and modeled

after pocket-sized editions of Ortelius's *Theatrum*. Translation of Mercator's Latin narrative into Dutch, French, German, and English created a still wider market for the thirty editions of the full-size Mercator-Hondius *Atlas* published between 1606 and 1641. The *Atlas minor* enjoyed an even longer run in the twenty-five editions Hondius and his successors published between 1607 and 1738. By 1700 numerous other publishers were issuing atlases, and the term was well established.

Perhaps the most compelling evidence of the *Atlas*'s endurance is its recent republication in CD format. In 2000 Octavo Digital Editions, an Oakland, California, firm headed by software designer John Warnock, issued a two-disc facsimile edition easily navigated with Adobe Acrobat Reader, the widely used electronic page-viewing application

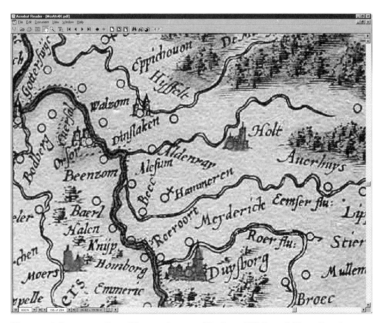

Figure 3.6 An excerpt from Mercator's map of Brabant, Jülich, and Cleve showing Duisburg (bottom center) and part of Cleve, as portrayed in the electronic edition of his 1595 *Atlas* published in color by Octavo, "Examine Disc," 155.

that Warnock helped develop. The "Read Disc" links Mercator's Latin text to an English translation and includes insightful commentary by map historian Robert Karrow. The "Examine Disc" consists of high-resolution scans of a copy in the Lessing J. Rosenwald Collection of the Library of Congress. Readers can turn the pages of the 1595 *Atlas,* peruse its maps, and zoom in for a detailed look at the mapmaker's conception of late sixteenth-century Europe. Figure 3.6, a close-up centered on Duisburg, where the mapmaker lived, illustrates the content and graphic detail, but the Octavo images, in full color, convey a fuller sense of the hand coloring and textured paper. Warnock's version also exemplifies the extension during the 1990s of the word *atlas* to include structured collections of viewable geographic data published on CDs or the Internet. Mercator's simple five-letter word apparently expresses the concept more effectively than the tedious synonym *geospatial database.*

Revealing Replicas

4

Few historic maps demonstrate as dramatically as Gerard Mercator's 1569 world map that size is both an asset and a liability. Printed on eighteen separate sheets and measuring 80 by 49 inches (202 by 124 cm) when fully assembled, its abundant space easily accommodates the detailed coasts and continents that proved a valuable source for smaller, less detailed world maps by Ortelius and Hondius, among others. Its suitability as a wall display, vulnerable to light and dirty fingers, partly explains the small number of surviving copies—a mere four complete sets if you include one lost in World War II—as well as its rare appearance as a facsimile illustration in books about old maps and cartographic history. When reduced to a black-and-white page-size halftone with noticeable lineations where its sheets meet, the famous chart becomes a disappointingly drab demonstration of its illustrious author's skill as a cartographer.

Several illustrators devised clean, book-friendly replicas by transcribing the map's key elements at a smaller, more manageable scale and adding labels describing its larger blocks of text. The resulting

map of a map (so to speak) not only affords a concise graphic sum-
mary of its prototype's geography but also allows for further reduction,
as in figure 4.1, which reproduces at a still smaller size the version
drafted by famed British engraver Emery Walker for the eleventh edition of *Encyclopaedia Britannica,* published in 1911. The warped
frame of the *Britannica* image, which measures 7 inches wide, suggests it might have been picked up photographically from an earlier illustration, perhaps in another publication. Despite this flaw, Walker's
diagram captures the essence of Mercator's grid, layout, continents,
and obvious belief in a southern continent (Antarctica) and northeast
and northwest passages across the Arctic. Similar renderings with English translations of legend labels illustrated a 1969 science article
commemorating the great map's four hundredth anniversary and a
1947 guidebook on map projection (*The Round Earth on Flat Paper*)
from the National Geographic Society. Although these small quasi fac-

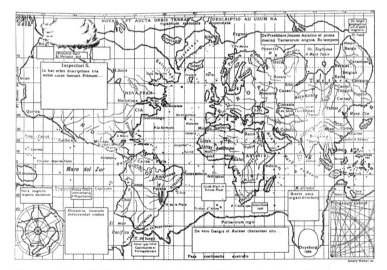

Figure 4.1 Emery Walker's reduced rendering of Mercator's 1569 world map enhanced the 1911 *Encyclopaedia Britannica*'s article on maps. From *Encyclopaedia Britannica,* 11th ed., 17:647.

similes lack the authenticity of Mercator's intricate engraving, they illustrate nicely the progressive poleward separation of his projected
parallels.

Full-size, eighteen-sheet facsimiles afford a more realistic impres-
sion of the famous mapmaker's attention to detail. The oldest is part
of *Les monuments de la géographie,* a collection of twenty-one facsimile maps published between 1842 and 1862 by Edmé-François Jomard
(1777–1862), a Parisian geographer who had served with Napoleon's
1798 Egyptian expedition. In 1828 Jomard founded what eventually
became the Bibliothèque Nationale's Department of Maps and Plans,
which owns one of the surviving copies. The municipal library in
Breslau (now the Polish city of Wroclaw) furnished the originals for
the Berlin Geographical Society's 1891 *Drei Karten von Gerhard Mercator,* which includes the 1569 world map as well as Mercator's 1554
fifteen-sheet map of Europe and his 1564 eight-sheet map of the
British Isles. Destroyed during World War II, the Breslau copy also
served as the prototype for a commemorative edition published in
1931 by the International Hydrographic Bureau, headquartered in
Monaco. The third known copy, in the Maritime Museum at Rotterdam, was reproduced in 1961 (a year before Mercator's 450th birth-
day) and distributed as a supplement to *Imago Mundi,* the principal
scholarly journal for historians of cartography. A fourth copy, at the
University Library of Basel, Switzerland, appeared in *Imago Mundi* in
1955 as a much-reduced 20 by 13 inch (51 by 33 cm) foldout.

Life-sized reproductions of Mercator's 1569 world map are not
widely available in libraries and map collections. A joint search of the
RLIN (Research Libraries Information Network) online catalog and
the OCLC (Online Computer Library Center) FirstSearch database,
which focus on university libraries in the United States, failed to turn
up a single copy of either the Jomard facsimile or its 1931 Monaco
counterpart, although the *National Union Catalog,* printed in the late
1960s, found ten of the former but none of the latter. Many American
libraries apparently did not bother to add older materials to their electronic catalogs. By contrast, my Web search uncovered twelve copies
of the Berlin Geographical Society's 1891 version in American librar-

ies—one more than the printed bibliography. Oddly, the Library of Congress (which apparently does not share all its holdings with RLIN or OCLC) owns both the Jomard and the Monaco editions but lacks the 1891 Berlin reprint. The 1961 Rotterdam facsimile fares much better: the Library of Congress has one, the *National Union Catalog* found nineteen more, and an RLIN/OCLC search uncovered another sixty copies in North American and Europe. The list would no doubt be longer had *Imago Mundi* automatically sent its Rotterdam "supplement" to all subscribers. Syracuse University, where I teach, has a run of *Imago Mundi* that starts with volume 1 (1935), but lacks a full-size facsimile of Mercator's 1569 world map. Yet the SUNY College at Cortland, forty miles south, has a copy, which I now know of thanks to FirstSearch.

If you think all "full-size" facsimiles look alike, think again. Although Gerard Mercator engraved eighteen separate plates and printed his 1569 world map on eighteen separate sheets of paper, the 1961 Rotterdam reprint consists of a large, atlas-like portfolio with fourteen huge pages that reformat the planet into geographically coherent chunks. A complex diagram (fig. 4.2) in the accompanying sixty-nine-page guide describes the scheme. The solid lines and large, bold numbers, 1 through 18, represent Mercator's original layout. The dashed lines represent the reformatted "sheets," also numbered 1 through 18. Some of the new sheets are smaller than others and share a page in the facsimile atlas with another sheet. Note too that Mercator's original plates were in portrait format (longer vertical axis), whereas the facsimile is in landscape format.

If this reformatting seems needlessly confusing, consider the inconvenient boundaries between Mercator's original plates. Anyone familiar with U.S. Geological Survey topographic maps knows the problem of quadrangle boundaries that invariably partition our area of interest, however small, among two, three, and sometime four map sheets. Mercator's 6 by 3 grid validates this cartographic variant of Murphy's Law with horizontal plate boundaries that separate England from Scotland and chop off the southern tip of Africa while vertical boundaries slice through east Africa and what is now the eastern

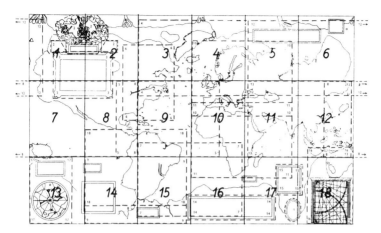

Figure 4.2 The diagram accompanying the 1961 Rotterdam facsimile relates the reformatted sheets (small numbers, dashed lines) to Mercator's original eighteen plates
(large numbers, solid lines). From "Gerard Mercator's Map of the World."

United States. Instead of retaining Mercator's original layout—a benefit only if you're obsessed with authenticity or eager to decorate a
wall—a benevolent editor chose to preserve the integrity of continents and other large regions. Thus a historian interested in medieval
Europe need only examine facsimile sheet 10, reformatted from plates
4, 5, 10, and 11 to include all of Europe (including Scandinavia) and
the entire Mediterranean coastline. Similarly, an Africanist can con-
centrate on sheet 11, reformatted from plates 10, 11, 16, and 17 to include all of Africa and the Arabian Peninsula. This useful redundancy
excuses the omission of tiny portions of Antarctica and northern
North America, where equivalent details are missing or inconsequen-
tial, as well as the museum's admission that "the size of the maps in
the reproduction is slightly reduced."

If you think it unseemly for anyone to muck around with the
great mapmaker's original layout, you'll not be surprised that the med-
dlesome editor of the Rotterdam copy was none other than Gerard
Mercator, who willingly sacrificed three prints of some plates in reformatting his huge wall map into an atlas of coherent pages, not arbi-

trary quadrangles. To provide a meaningful geographic scope for each sheet, Mercator cut out appropriate portions of prints covering the region in question and mounted them on large, folio-size sheets of paper. Although the guide ignores the composition and condition of Mercator's adhesive, lines between adjoining pieces are readily apparent in the 1961 reprint. Repackaged as a portfolio atlas, the Rotterdam copy is notably different from its intact, eighteen-plate cousins in Basel and Paris, and the Maritime Museum owns the only copy so arranged.

A handwritten letter from Mercator found with the atlas indicates he assembled it around 1578 at the request of Werner von Gymnich, a wealthy patron. Sometime later a von Gymnich married a von Mirbach, and the atlas moved to the Mirbach estate at Harff Castle, near Cologne, where it evaded the public gaze until the late nineteenth century. Discovered in the course of an inventory of the Mirbach family library, the atlas was exhibited briefly in 1894, written up in a Frankfurt newspaper in 1902, and mentioned in academic articles in 1911 and 1930. Pieced together from printed map sheets, it earned recognition in Leo Bagrow's seminal *History of Cartography* as "the first printed sea-atlas."

Priceless objects become disposable assets during hard times. In 1932, amid the economic and political turmoil that eased the Nazi rise to power, the atlas appeared in the catalog of an auction house in Lucerne and caught the attention of the Maritime Museum's director, J. W. van Nouhuys. Eager to acquire a rare work of immense importance to cartography and navigation, van Nouhuys decided to attend the sale. On the way he stopped at Basel to examine the copy in the university library. His hopes rose when an overly optimistic auctioneer opened the bidding at 6,000 Swiss francs (3,000 guilders), raised the price twice, and failed to find a buyer. Convinced he had a chance, van Nouhuys returned home, found two wealthy contributors each willing to match the museum's 1,000 guilders, and mailed in the winning bid. The museum got its atlas.

On at least one other occasion Mercator repackaged a wall map as an atlas. The evidence is a bound collation of regionally reconfigured cutouts from his maps of Europe (1554) and the British Isles (1564) as well as portions of the 1569 world map and thirty additional sheets from his friend Abraham Ortelius's 1570 world atlas. Dutch schoolmaster Thomas Varekamp, who discovered it serendipitously in a Belgian used-book shop in 1967, reckons that Mercator assembled the atlas in 1571 for his patron Werner von Gymnich, who undertook a lengthy tour of Europe the following year. As with the Rotterdam sea atlas, Mercator willingly cannibalized his wall maps to get the right framing. The nine sheets he assembled by cutting up four copies of his European wall map are especially rare because the last known intact copy was destroyed in 1945, during the siege of Breslau, along with the municipal library's copy of the 1569 world map. A pair of manuscript maps of northern Italy—the only surviving examples of maps hand-labeled by Mercator—makes the atlas triply unique. Robert Karrow, who considers it "the most important Mercator discovery of the twentieth century," once lamented the atlas's purchase in 1979 by the British Rail Pension Fund, which persistently refused scholars' requests for close inspection. But not any longer: a 1997 grant from the Heritage Lottery Fund enabled the British Library to purchase what is now known as *Mercator's Atlas of Europe* and endorse a facsimile reprint, which appeared a year later.

If you want your own copy of the 1569 world map and are willing to settle for less than full size, Respree.com, a Los Angeles dealer in posters and reproductions, advertises two poster versions at its Web site, one 24 by 31 inches and the other 37 by 54 inches. Reproduced in color from an unidentified original, Respree's posters make attractive wall decorations but lack the detail of Mercator's lines and labels, better captured by a full-size black-and-white facsimile.

If you crave fine details, don't mind low-resolution scanned images, and can appreciate a well-illustrated online celebration of Mercator's projection, all in German, check out the "Ad Usum Navigantium" page of Wilhelm Krücken's Web site, "Ad maiorem Gerardi Mercatoris gloriam" (www.wilhelmkruecken.de). In addition to an incisive exploration of the projection's mathematical properties, Krücken provides individual screen-size images of all eighteen plates. Although the map's Latin inscriptions are barely discernible, its geography and artistry are clearly apparent. Anyone with a high-speed Internet connection can move from plate to plate more readily than a library patron handling (carefully, I hope) eighteen oversize map sheets. It was during a cursory interactive examination of Krücken's facsimile that I first understood the great mapmaker's appreciation of modest amounts of redundancy. Unlike contemporary cartographers who give you the whole world only once, Mercator cleverly extended his left-hand plates a few degrees west of his 180th meridian and his right-hand plates a few degrees east to provide dual, alternative images (fig. 4.3) of what he considered the most likely position of the north magnetic pole.

For scholars concerned with a map's lines and labels, an accessible black-and-white facsimile is often more valuable than a rare hand-colored print ensconced in a distant library. That's clearly the view of

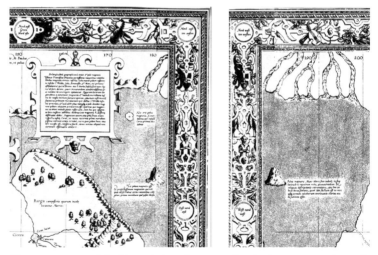

Figure 4.3 Mercator's double rendering of a steep, rocky island representing the north magnetic pole in the far northeast (left) and far northwest (right) corners of his 1569 world map. Cropped and reduced for convenience and legibility from Wilhelm Krücken's online facsimile. From Krücken, "Ad maiorem Gerardi Mercatoris gloriam," http://wilhelmkreuken.de/adusum/11.htm and http://wilhelmkreuken.de/adusum/16.htm.

Swedish scientist-explorer Adolf Erik Nordenskiöld (1832–1901), who inspired map historians with his *Facsimile Atlas to the Early History of Cartography,* published in 1889 and reprinted in paperback in 1973. Unable to include Mercator's 1569 world map because of its huge size, he reported an analysis of its progressively spaced parallels. One column of a table contains distances from the equator calculated for a spherical earth according to the mathematical formula for the Mercator projection, and a second column lists the corresponding distances as measured on the Jomard reprint. The two columns are nearly identical up through 20°, but a discrepancy apparent at 30° grows progressively larger with increased latitude. In attributing this discrepancy to "the imperfection of the mathematical resources of the map-constructors in the middle of the 16th century," Nordenskiöld suggests that Mercator employed a mathematical approximation that should yield a closer correspondence between the calculated and measured distances. Remaining differences, he argues, "can be explained by engraving-errors or by stretchings in the paper"—a persistent source of uncertainty when working with facsimiles of old maps.

Paper shrinkage is less troublesome in exploring Mercator's reliance on other mapmakers. In announcing the 1931 Monaco reprint by the International Hydrographic Bureau, Britain's *Geographical Journal* noted several dubious debts to Ptolemy, including a Niger River that connects with the Nile. Bert van 't Hoff, who prepared the guide accompanying the 1961 Rotterdam facsimile, listed other prominent influences, which are equally apparent on original copies and reprints. Especially noteworthy is Abraham Ortelius's 1564 world map, on which inscriptions nearly identical to their counterparts on Mercator's 1569 world map suggest that Mercator and Ortelius exchanged information or consulted the same sources. Impressed with an original print of the Ortelius map in the library at Basel, van 't Hoff observed that "this beautiful and remarkable map deserves to be reproduced [and] also compared in detail with Mercator's map." He also recommended looking at Diego Gutierez's 1562 map of South America (fig. 4.4, left), the likely source of Mercator's erroneous westward diversion of the coastline for what is now southern Chile (see fig. 4.1).

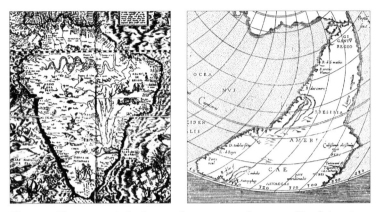

Figure 4.4 The comparatively plump and erroneous southern part of South America on Diego Gutierez's 1562 map of the continent (left) is apparent on Gerard Mercator's 1569 world map (see figure 4.1). From LeGear, "Sixteenth-Century Maps," facing 18. The thinner, more accurate rendering on the great mapmaker's 1538 world map (right), inverts the continent. From Nordenskiöld, *Facsimile Atlas,* facing 91.

The straighter, more accurate rendering of the Pacific shoreline on Mercator's 1538 world map (fig. 4.4, right) confirms the adage that new is not always better.

One map certain to warrant a facsimile reprint—if it's ever found—is a larger version of Erhard Etzlaub's small maps of Europe and North Africa on what looks like a Mercator projection (see fig. 1.7). As I noted in chapter 1, two tiny maps produced in Nuremberg in 1511 and 1513 exhibit progressively spaced parallels, which suggest a deliberate attempt to straighten out rhumb lines. Facsimiles of these maps, each engraved on the cover of a portable sundial, have made map historians wonder what Etzlaub was up to and whether his tiny maps had influenced Mercator. In 1918, in a short note titled "Who Originated Mercator's Projection?" the *Geographical Journal* reported the opposing views of two German professors, Drecker and Hammer. Convinced that Etzlaub's maps were not flukes, Drecker believed the Nuremberg instrument maker had merely reproduced miniature versions of a much larger map, yet to be discovered. Hammer, who questioned Etzlaub's understanding of principles underlying the Mercator

projection, wondered why no sixteenth-century geographer or mathematician had ever discussed his alleged innovation.

Curious about current views among European map historians, I
put the question of Etzlaub's influence on Mercator to Ingrid Kretschmer, who teaches the history of cartography at the University of
Vienna. She began by noting that the University of Duisburg—the
duke's idea was eventually fulfilled—celebrated the four hundredth
anniversary of the great mapmaker's death in 1594 with three sym-
posia, in 1992, 1993, and 1994. The series stimulated an intense reex-
amination of the Mercator's work, including a careful comparison of
his and Etzlaub's maps by mathematician Wilhelm Krücken, who
maintains the Mercator Web site mentioned earlier. According to
Kretschmer, a detailed examination of graticules convinced Krücken
that the two cartographers had applied different principles. Renewed
interest in Etzlaub's influence failed to uncover a larger version of his
tiny sundial maps—the several, larger scale road maps he published
are all laid out on an equirectangular grid. More telling is Etzlaub's apparent failure to tout his accomplishment in a published article or private correspondence. In Kretschmer's opinion, it is "rather unlikely
that a famous instrument maker and cartographer like Erhard Etzlaub
would not have mentioned [his development of] a new projection."

Although Etzlaub's influence on Mercator remains a mystery, the
great mapmaker might well have been inspired by the work of Pedro
Nuñes (1502–78), a Portuguese astronomer and mathematician who
described loxodromic spirals in 1537. In pointing out that a direct
course is usually not the most easily followed course, Nuñes criticized
globe makers for confusing great circles (orthodromes), which are easily delineated on a globe by a taut thread, with rhumb lines (loxodromes). Unless aligned with a meridian or parallel, a rhumb line is a
comparatively complex corkscrew curve.

Mercator knew about loxodromic spirals as early as 1541, when he
included a multitude of these curved lines of constant direction on his
famous terrestrial globe (fig. 4.5). How he did this invites speculation:
Mercator never described his method, and scholars have yet to uncover an earlier prototype. Dutch map historian Johannes Keuning

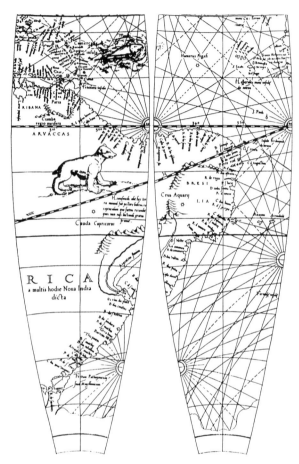

Figure 4.5 A facsimile of a pair of unmounted gores printed for Mercator's 1541 terrestrial globe shows curved loxodromes converging at thirty-two-point compass roses.
From Kretschmer, "Mercators Bedeutung," 165, fig. 5.

concluded that Mercator simply drew his loxodromes graphically,
"with the aid of metallic loxodromic triangles, made by himself." Using thin metal triangles or templates crafted to ensure constant bearings seems both obvious and ingenious. I'm surprised that Keuning,
after proposing a graphic, ad hoc solution for placing loxodromes on a

globe, did not dispute Nordenskiöld's assumption that Mercator used a mathematical approximation in laying out his 1569 world map.

Although mathematics provides a theoretical foundation for map projection, not all solutions are numerical. It's easy to see how a mapmaker sufficiently clever to delineate curved loxodromes on globe gores could have straightened them out graphically on a flat map. Quite simple, in fact, according to Robert Karrow, who summarizes a likely Mercator recipe in a single sentence: "By following his curved rhumb lines and noting the longitude at which these lines crossed the various parallels, then transferring these coordinates to flat paper using straight rhumb lines with the same bearing, he would have obtained the basic framework of his projection." As William Warntz and Peter Wolff point out in *Breakthroughs in Geography,* Mercator's genius lay in believing a solution existed. It might well have happened, they note, "that the requirement that rhumb lines should be straight lines could not be satisfied on *any* chart or map."

If copying is a clue, other mapmakers appreciated Mercator's genius less readily than his scholarship. Although no one adopted his projection for nearly three decades, his 1569 world map was a key source for his friend Abraham Ortelius's influential *Theatrum Orbis terrarum,* published in 1570. According to map historian Peter Meurer, Ortelius based eight of his plates on Mercator's map. A few writers have interpreted these similarities as blatant plagiarism, but most note that Mercator and Ortelius exchanged information and consulted similar sources. Ortelius was not the only mapmaker to rely on Mercator's geography. Features and place names from the 1569 wall map (but not its projection) are readily apparent on the world map accompanying the *Polyglot Bible,* published in Antwerp by Benito Arias Montanus in 1572, and the "Planisphere," a world map published in Antwerp and Amsterdam by Petrus Plancius (1552–1622) in 1592.

Mercator's 1569 world map proved a convenient source for his son and grandsons, who compiled continental maps for the posthumous 1595 edition of his famous *Atlas,* which also includes a world map his son Rumold had published separately in 1587. Laid out on a pair of

hemispherical projections, Rumold's map is a smaller-scale generalization of his father's much larger 1569 chart. Paradoxically, not a single
map in the epic 1595 *Atlas* is on a Mercator projection.

Another paradox is the full-size 1574 woodcut reprint of the 1569
world map by Bernard van den Putte, an Antwerp engraver. With the
apparent approval of Mercator, van den Putte reengraved all eighteen
plates by cutting out non-inked areas on blocks of wood analogous to
massive rubber stamps. Although coastlines and other features left
standing on a woodblock are less elegant than lines cut into a copper
plate, a wood engraver could add place names or descriptive text by
merely cutting a rectangle into the wood and inserting pieces of metal
type. That only a single sheet survives suggests that van den Putte's
version was less commercially successful than Mercator's copperplate
edition. Even so, this mechanical facsimile, which acknowledged Mercator's authorship, contributed to the projection's growing prominence and could have inspired one or more of three substantially

Figure 4.6 Engraved in 1594, this world map on a Mercator projection enhanced
Matthew Quad's *Geographical Handbook*. From Nordenskiöld, *Facsimile Atlas*, pl. 49 (3).

smaller Mercator maps published around 1595. Nordenskiöld included one of them in his *Facsimile Atlas*: a 9 by 12 inch (22 by 32 cm) copperplate engraving from Matthew Quad's *Geographic Handbook,* published in Cologne between 1594 and 1608. Although Quad's map (fig. 4.6) lacks a graticule, its origin is plainly apparent in both its title (". . . *ad imitationem universalis Gerhardi Mercatoris*") and the telltale shapes of its continents.

If multiple maps by diverse authors are a reliable indicator, Mercator's projection became the cartographic expression of a hot idea in the late 1590s, when Jodocus Hondius and Edward Wright offered their own versions of Mercator's world. Hondius's map predates Wright's, but as the next chapter points out, the Dutch cartographer relied heavily on the English mathematician, who developed a mathematical description as well as tables showing how to position the parallels. Although Mercator demonstrated the projection's look and use, Wright made the secret of its construction readily available to other mapmakers.

The Wright Approach

Edward Wright was a mathematician, not a mapmaker. Born in 1561 in the village of Garveston, about one hundred miles northeast of London, he attended Gonville and Caius College at Cambridge, where he received a bachelor of arts degree in 1581 and a master of arts three years later. In 1587 a research fellowship allowed him to focus on mathematical cosmography and its use in navigation. In 1589 a seven-month leave to help the Earl of Cumberland plunder Spanish shipping in the Azores provided practical experience at sea. Appalled by mariners' misuse of almanacs, charts, and navigation instruments, Wright undertook a mathematical critique of contemporary navigation. His search for new solutions to old problems included a sea chart with straight-line loxodromes: the map Mercator had demonstrated but never explained.

Wright compared the projection to an inflatable globe inside a glass cylinder. Imagine a spherical bladder, he suggested, with meridians, parallels, and a selection of rhumb lines inscribed on its surface. Inflate the sphere initially so that its axis aligns with the axis of the cylinder and its equator just touches the glass. This step establishes the equator as the standard line, with the same scale on both globe and cylinder. Then slowly inflate the bladder so that all rhumb lines remain straight and the stretching at every point is the same in all directions—the angle-preserving condition now known as conformality. Although Wright's mythical model requires an infinitely expansive bladder of extreme flexibility, it describes perfectly the transformation of a globe into Mercator's conformal cylindrical projection: the parallels grow farther and farther apart as the bladder inflates, but because the cylinder is open ended, the poles never touch the map.

To describe the growing separation of the map's successive parallels Wright worked up a table with three columns. The first two list the degrees and minutes of latitude for parallels spaced ten minutes apart on the sphere, and the third reports the parallel's projected distance from the equator. Because the northern and southern hemispheres have identical grids, the table runs from the equator at $0°$ to a generic pole at $90°$, and because a degree contains sixty minutes, an interval of ten minutes divides each half meridian into 540 (90×6) "meridional parts." To simplify the calculations, Wright set to 100 the distance encompassed by an arc of ten minutes at the equator. With minimal distortion near the equator, the parallels for $0° 10'$ and $0° 20'$ plot at 100 and 200 distance units, respectively. Because the table's third column has no decimal places, the slowly growing separation of parallels is not apparent until the sixteenth meridional part positions the parallel for $2° 40'$ at 1,601—up 101 units (rather than 100) from the parallel for $2° 30'$ at 1,500. Vertical stretching becomes only slightly more apparent at $15° 00'$, which plots at 9,104—only 103 distance units away from $14° 50'$, which plots at 9,001. Separations increase, and in its final rows the table locates the parallels for $89° 40'$ and $89° 50'$ at 201,513 and 226,223, respectively, and describes the polar parallel of $90° 00'$ as "Infinite." With Wright's "Table for the true dividing of the meridians in the Sea Chart," any mapmaker or sailor could easily lay out a Mercator grid.

Wright used at least three decimal places in his calculations, but omitted them from the abridged table in the first edition of *Certaine*

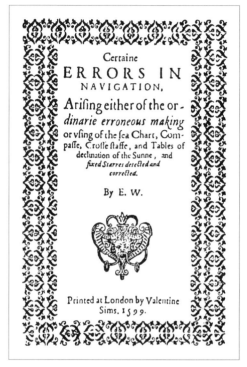

Figure 5.1 The title page of the first edition of Wright's treatise.

Errors in Navigation (fig. 5.1), published in 1599, in order "not at this time to trouble [chartmakers and navigators] with more than thought to be of use." Another concern might have been his publisher's bottom line: the condensed table with a ten-minute interval occupies a mere six pages in the 1599 edition, whereas the complete table, with a one-minute interval and smaller type, consumes twenty-three pages in the second edition, published in 1610. Of little direct use to most readers, the added precision of 5,400 (90 × 60) small meridional parts minimized cumulative error.

However tedious, Wright's calculations are straightforward. The map's rectangular grid, which stretches the parallels to equal the equator in length, compensates for this increasing horizontal exaggeration

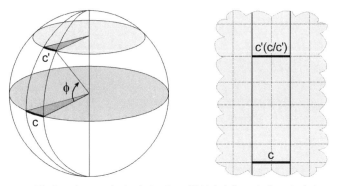

Figure 5.2 The key elements in the derivation of Wright's formula for calculating projected positions of parallels. At latitude ϕ the arc distance c' between two meridians (left) is measurably less than c, its corresponding length at the equator. Because the meridians on the Mercator projection do not converge (right), the projected arc at latitude ϕ is exaggerated by a factor of c/c'.

by shifting the parallels farther apart vertically. The left part of figure 5.2 describes key elements in the calculation: a pair of meridians divide the equator and a parallel at latitude ϕ into sections with lengths c and c', respectively. Note that on the globe c' becomes progressively smaller than c with increasing latitude. Because the meridians on the map (right side of fig. 5.2) cannot converge, the mapped arc at latitude ϕ is stretched horizontally by a factor of c/c'. At 60°, where the full circumference of the parallel on the globe is half the length of the equator, the stretching ratio c/c' equals 2.0. Farther poleward, as the latitude approaches 90°, c' shrinks to zero and the stretching factor approaches infinity. Near the equator, though, east–west stretching is comparatively minor and the ratio is only marginally greater than 1.

Trigonometry conveniently enters the picture at this point because c/c' is the secant of angle ϕ. (In trigonometry the secant of an angle in a right triangle is the ratio of the length of the hypotenuse to the length of the adjacent side.) By consulting a table of secants, readily available to any late sixteenth-century university mathematician, Wright could look up the stretching factor for any latitude.

The crux of Wright's method is a cumulative vertical adjustment for horizontal exaggeration. Figure 5.3 describes the process. A pair of meridians one minute apart define the east and west sides of a series of quadrangles covering one minute of latitude or longitude on all sides and extending upward from the equator. Stacked vertically, the quadrangles have curved sides on the sphere but plot as rectangles on the projection. Because the map's meridians cannot converge, the first quadrangle, which covers distance d along the equator, must be slightly taller to compensate for east–west stretching along its upper edge, which is proportional to the secant of its latitude. Thus its upper edge, defined by the parallel at $0°$ $1'$, is d times the secant of $1'$, which is written as $d \sec 1'$. Similarly, the height on the map of the second quadrangle, ever so slightly taller, is $d \sec 2'$, and the height of the quadrangle immediately above is $d \sec 3'$. As the diagram shows, the vertical distance from the equator to the parallel at $0°$ $3'$ is the sum of the heights of these three rectangles. More generally, the map distance y from the equator to the parallel at latitude ϕ can be computed as

$$y = d \left(\sec 1' + \sec 2' + \sec 3' + \cdots + \sec \phi\right).$$

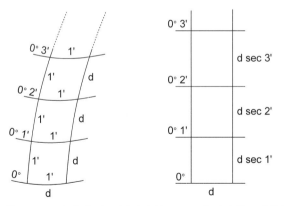

Figure 5.3 To compensate for horizontal stretching, one-minute ($1'$) quadrilaterals on the globe (left) plot on the map as vertically inflated rectangles (right). Because this vertical stretching increases with latitude, parallels with a one-minute separation on the globe grow progressively farther apart on the map. For clarity, the right-hand diagram flagrantly overstates vertical stretching.

How accurate is Wright's table? To find out, I wrote a short computer program—something cartographers rarely do these days, now that commercial software handles most mapping tasks. Table 5.1 compares my results with Wright's abridged and complete tables, published in 1599 and 1610, respectively. Although the differences are more apparent for higher latitudes, where cumulative error should be most noticeable, the numbers are remarkably close. To put these discrepancies in perspective, I calculated the error (assuming my computer is reliable) for a world Mercator map three feet wide. At this scale the greatest discrepancy, a mere 2.366 at 80°, represents a nearly infinitesimal 0.00039 inches on the map—well within the tolerance of the most precise automatic plotters. It's hard not to be both amazed and impressed.

The calculations no doubt impressed mapmaker-engraver Jodocus Hondius, who was in London in the early 1590s, sitting out the Netherlands' version of the Spanish Inquisition. Hondius heard of Wright's work and borrowed a draft manuscript for a brief period after agreeing not to publish any of its contents without permission. But an accurate table of meridional parts was too great a temptation for the Dutch mapmaker, who drew on the English mathematician's labors for several regional maps as well as a world map he published in Am-

Table 5.1 Comparison of Edward Wright's abridged (1599) and complete (1610) tables of meridional parts with values computed electronically

Latitude	Wright (1599)	Wright (1610)	Computer (recent)
0°	0	0.0	0.0
10°	6,030	6,030.475	6,030.773
20°	12,251	12,251.292	12,251.772
30°	18,884	18,883.768	18,884.528
40°	26,228	26,227.559	26,228.430
50°	34,746	34,746.045	34,747.508
60°	45,277	45,277.106	45,278.680
70°	59,667	59,666.811	59,668.803
80°	83,773	83,773.416	83,775.782

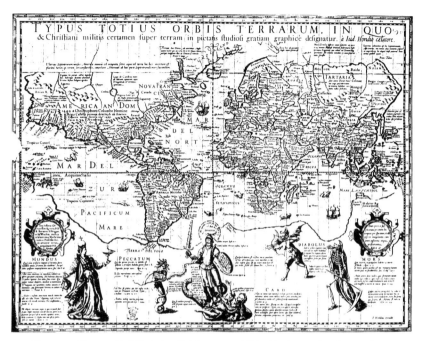

Figure 5.4 The so-called Christian-Knight map, published in Amsterdam in 1597 by Jodocus Hondius, who used Wright's calculations without permission in laying out the projection. The original measures 15 by 19 inches (37 by 48 cm). From Shirley, *Mapping of the World*, 218, pl. 161.

sterdam in 1597 (fig. 5.4). No less apparent than the progressive pole-ward spacing of the map's parallels is the allegorical engraving of a Christian knight battling Sin, the Flesh, and the Devil. Hondius was mute about how he laid out the grid but dedicated the map, in Latin, to "Ed. Wrichto" and two other Englishmen.

Wright was outraged. In the preface to *Certaine Errors,* he quoted a letter in which Hondius had offered a vague apology: "I hear that you are somewhat offended with me because I have taken those few things out of your hand-written book. . . . Truly I told all my friends plainly that you are the author thereof, and I tell them so still." In what historian Lawrence Wroth termed "the most inept rationalization a plagiarist ever made," Hondius pleaded, "I was purposed to have set this forth under your name, but I feared that you would be displeased therewith because I have but rudely translated it into Latin." Neither moved nor mollified, Wright spared no sarcasm in condemning his former friend's deceit and greed: "But how well and honestly he [honored his agreement], grounded upon faith and credit, the world may now see: and how thankful he hath been to me for that which hath been so profitable and gainful unto himself, as may appear by so common sale of his maps of the world, and of Europe, Asia, Africa, and America (all which had been yet unhatched, had he not learned the right way to lay the groundwork of them out of this book) I myself know too well. But let him go as he is."

The Christian-Knight map (as map historians call it) was not the only premature publication of Wright's results. His table of meridional parts appeared in print in 1594, in mathematician-navigator Thomas Blundeville's *Exercises for Young Gentlemen,* and again, three years later, in Sir William Barlow's *The Navigator's Supply.* Although both authors had obtained Wright's permission, only Blundeville acknowledged him by name. Barlow vaguely credited "a friend of mine of like profession." More devious was Abraham Kendall, a navigator with Sir Robert Dudley's 1594 expedition to Guiana and Trinidad. Kendall borrowed a draft of Wright's manuscript, made a longhand copy without permission, and carried it with him on Sir Francis Drake's 1595 expedition to the West Indies. Whatever his intentions, Kendall died off Porto Bello, and the manuscript found its way back to London, where someone, thinking it original scholarship, sent it to the Earl of Cumberland, who immediately recognized the work of his former hydrographer. According to maritime historian David Waters, two brushes with plagiarism—at the hands of Kendall and Hondius—convinced Wright to publish his book, which he gratefully dedicated to the Earl.

Although Wright was neither the engraver nor the publisher, historians credit him with the two-sheet world map (fig. 5.5) prepared for the second volume of geographer-navigator Richard Hakluyt's *Principal Navigations, Voyages, Trafiques and Discoveries of the English Na-*

tion, published in 1599. Sometimes called the Wright-Molyneux map because Wright laid out the graticule and transferred features from Emery Molyneux's 1592 terrestrial globe, the chart reflects recent discoveries by explorers like John Davis. Conspicuously absent are the northern Pacific coasts, the vast southern continent, and other questionable features that populate unexplored regions on most late-sixteenth-century world maps. Hungry for accurate information, Wright and his co-compilers consulted the latest Dutch, Portuguese, and Spanish charts and amassed a total of 1,209 place names, mostly coastal. Acclaimed by the English intelligentsia for correcting numerous inaccuracies on existing charts, the map's fame is affirmed in act 3, scene 2 of William Shakespeare's *Twelfth Night,* in the line "He does smile his face into more lines than is in the new map with the augmentation of the Indies."

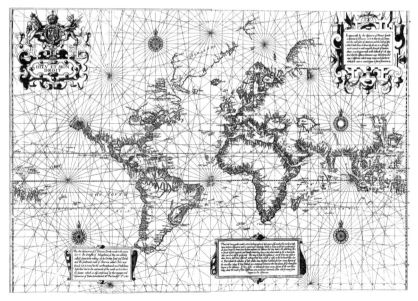

Figure 5.5 The Wright-Molyneux world map of 1599, included with some copies of volume 2 of Richard Hakluyt's *Principal Navigations.* The original measures 17 by 25 inches (43 by 64 cm). From Nordenskiöld, *Facsimile Atlas,* pl. 150.

Wright died in 1615, knowing he had made an important contribution to navigation and cartography. Although his principles and calculations promoted a wider use of Mercator's projection—which a few logrolling British historians proposed calling the Wright-Mercator projection—it's now clear that another English mathematician, Thomas Harriot (1560–1621), had begun to address the problem of meridional parts around 1589, about the same time as Wright. What's more, Harriot's solution is cleaner and more mathematically elegant insofar as he had progressed from merely adding up secants, as Wright had done, to a logarithmic tangents formula that affords a more exact and direct solution. We know this because Harriot left behind a massive collection of unpublished drawings, tables, notes, and manuscripts—over ten thousand pages worth, according to the *Dictionary of Scientific Biography,* which attributes his aversion to publishing to "adverse external circumstances, procrastination, and his reluctance to publish a tract when he thought that further work might improve it." A brilliant scholar with a profound understanding of astronomy and physics as well as mathematics, Harriot is the epitome of the perfectionist academic who rarely publishes.

Harriot's solution anticipated the serendipitous discovery of another English mathematician, Henry Bond (ca. 1600–1678), who around 1645 noticed a surprising correspondence between Wright's table of meridional parts and a table of logarithms of tangents published in 1620 by Edmund Gunter (1581–1626). It wasn't a direct correspondence—Bond had to reorganize Gunter's table to show logarithmic tangents of $(45° + \phi/2)$, where ϕ is latitude—but once the numbers were rearranged, an exact match suggested strongly that the distance y from the equator of the parallel at latitude ϕ on a Mercator projection could be computed as

$$y = R \ln \tan (45° + \phi/2),$$

where R is the radius of a globe that defines the projection's scale and ln specifies a natural (or Napierian) logarithm. Bond's insight is important for two reasons. First, because the equation is not based on a succession of sums, it promotes a more straightforward, less error-prone calculation of projected coordinates using either a computer or a table of logarithmic tangents. (A moot point, perhaps, if Wright's table is at hand.) Second, and more important, as an equation readily manipulated using algebra and calculus, Bond's formula fosters a detailed mathematical examination of the projection's geometric distortion.

I looked in vain for a copy of the 1645 edition of Richard Norwood's *Epitome of Navigation,* in which Bond, who was its editor at the time, first published his observation. But no less than the eminent Edmund Halley confirmed Bond's discovery in a 1696 essay in the *Philosophical Transactions of the Royal Society of London.* Halley titled his article "An Early Demonstration of the Analogy of the Logarithmick [*sic*] Tangents to the Meridian Line or Sum of the Secants." After crediting "our Worthy Countryman Mr. *Edward Wright*" with a valuable table "to be met with in most Books treating of Navigation, computed with sufficient exactness for the purpose," he turned to the subject of his essay in noting, "It was first discovered by chance, and as far as I can learn, first published by Mr. *Henry Bond,* as an addition to *Norwoods Epitome of Navigation,* about 50 Years since, that the *Meridian Line was Analogous to a Scale of Logarithmick Tangents of half the Complements of the Latitudes.*" Halley's article is important because he not only validates the Bond legend but also substantiates Wright's claim to priority. Like Wright, Halley was innocently ignorant of Harriot's unpublished solution.

Like most mathematicians I know, Halley was less concerned with the proposition's history than with its proof. An earlier proof, by James Gregory (1638–75), was hardly elegant, or as Halley saw it, "not without a long train of Consequences and Complications of Proportions, whereby the evidence of the Demonstration is in a great measure lost, and the Reader wearied before attaining it." And while subsequent attempts strayed from the point of Bond's discovery, Halley's own demonstration, the focus of his essay, was simple, on target, and probably original, as he boldly asserts in a remarkably candid and irresistibly quotable disclaimer:

Wherefore having attained, as I conceive, a very facile and natural demonstration of the said Analogy, and having found out the Rule for exhibiting the *difference of Meridional parts,* between any two parallels of Latitude, without finding both the Numbers whereof they are the difference: I hope I may be entitled to share in the emprovements of this useful part of Geometry. Desiring no other favour of some Mathematical Pretenders, than that they think fit to be so just, as neither to attribute my desire to please the Honourable Royal Society in these Exercises, to any kind of *Vanity* or Love of Applause in me, (who too well know how very few these things oblige, and how small reward they procure) nor yet to complain, *coram non judice,* that I arrogate to my self the Inventions of others, and upon that pretext to depreciate what I do, unless at the same time, they can produce the Author I wrong, to prove their assertions. Such disingenuity as I have always most carefully avoided, so I with not too much experience of it in the very same persons, who make it their business to detract from that little share of Reputation I have in these things.

If Thomas Harriot had been as eager to publish, Edward Wright might be no better known today than Abraham Kendall or Henry Bond.

In a self-esteem contest, Halley could not hold a candle to Johann Heinrich Lambert (1728–77). During an interview for membership in the Prussian Academy of Sciences, Frederick the Great asked Lambert to name the science in which he was most proficient. Without hesitation, the candidate calmly answered, "All." Hardly an overstatement, though, for a genius whose contributions encompass mathematics, physics, astronomy, philosophy, and cartography. Born in Alsace to poor German parents and largely self-educated, Lambert worked as a clerk, secretary, and tutor before moving to Berlin in 1764. According to the *Dictionary of Scientific Biography,* his appointment to the Academy was delayed a year because of "his strange appearance and behavior." Lambert was openly religious, perhaps obnoxiously so, and he had an exceptionally high forehead, highlighted in the intriguing portrait (fig. 5.6) that decorates nearly every account of his life and work.

Figure 5.6 A lithographic engraving of J.-H. Lambert. From
Maurer, "Johan Heinrich Lambert," facing 70.

I suspect, though, that the engraver, working from sketches decades
after his eminent subject's demise, exercised a bit of artistic license in
endorsing popular ideas about superior intelligence and cranial ca-
pacity.

Lambert's contributions to cartography include seven different
map projections as well as an illuminating mathematical analysis of
conformality. In addition to using calculus to derive Bond's analytical
formula for the Mercator projection, he demonstrated that the Mercator map is a "special case" in a family of conformal projections with
polar and conic versions. As figure 5.7 illustrates, the cylinder and the

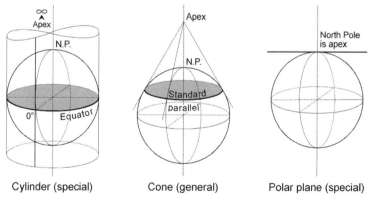

Figure 5.7 Lambert recognized the cylinder and polar plane as special cases of the
cone.

plane are extreme forms of a cone tangent to the sphere along a "standard parallel." Positioning the apex at infinity converts the cone to a cylinder, with the standard parallel at the equator. Putting the apex on the North Pole flattens the cone to a plane and shrinks the standard parallel (at 90°) to a point. If the projections are conformal, the cylindrical case is the Mercator, the planar case is the polar stereographic (in use since about 150 BC), and all intermediate cases are instances of the Lambert conformal conic projection, presented in 1772.

Lambert's insight stimulated further work on map projection by three of the era's greatest mathematicians, Euler, Lagrange, and Gauss. For me, though, the next most decisive contributor is an otherwise obscure Paris mathematics teacher, Nicolas Auguste Tissot (pronounced "tea-so"), who devised an analytical description of map distortion. (I searched for a biography or obituary, but found nothing.) Tissot's monograph *Mémoire sur la représentation des surfaces et les projections des cartes géographiques,* published in 1881, focuses on "the indicatrix," a simple device for describing distortion of angles and shape. Picture a globe with many small circles—infinitesimally small, in theory—all the same size. On conformal projections, which do not distort angles, the tiny circles remain circles but vary in area—as Mercator's map demonstrates, conformal projections suffer severe areal distortion in zones far from a standard line. By contrast, on projections that are not conformal, compression and stretching deform most circles into ellipses as shown in the indicatrix in figure 5.8. In

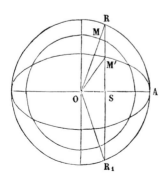

Figure 5.8 The indicatrix. From Tissot, *Mémoire,* 15.

this example, point M on the circle corresponds to point M′ on the ellipse, which reduces the angle ROR_1 on the globe by an amount equal to twice the angle MOM′. Using calculus and his indicatrix, Tissot calculated areal distortion or maximum angular distortion at grid intersections for a variety of projections, including Mercator's. In the next century his formulas helped analytical cartographers design customized projections that minimize distortion for specific regions.

As a graphic device for evaluating map projections, Tissot's indicatrix is unrivaled. Anyone who grasps the notion of a network of small, uniform circles on the globe can easily compare areal distortion on the Mercator projection with angular distortion on the Peters map. On the Mercator map (fig. 5.9, left) small circles grow ever larger with

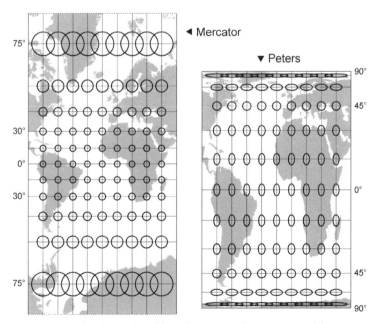

Figure 5.9 Tissot's indicatrix for a fifteen-degree graticule compares areal distortion on the Mercator projection (left) with angular distortion on the Peters projection (right).

increasing distance from the equator—the price of preserving angles and loxodromes on a rectangular projection. An altogether different trade-off arises with the Peters projection (fig. 5.9, right), on which perfect shape at 45° N and S gives way to severe north–south stretching in the tropics and an equally troubling east–west wrenching around 75°, where severely deformed ellipses overlap. In neither case would the indicatrix be plotted at the poles, where east–west scale is indefinitely large. In chapter 9, Tissot's clever device supports an insightful appraisal of promising substitutes for the Peters projection, such as those of Robinson and Goode.

Given Tissot's contribution to the visual evaluation of map projections, it's ironic that his treatise contains very few diagrams and no maps. Hardly surprising, though: mathematicians like Lambert and Tissot were numerical theoreticians, not mapmakers. Proficient in successfully attacking important cartographic problems analytically, they had little concern for the practical implications of their work.

Travelers' Aide

Edward Wright's denunciation of "the ordinarie erroneous making or using of the sea chart" had little immediate impact on mariners, who resisted his "new map" well into the eighteenth century. Most navigators trusted tradition more than science, and the plane chart, though not perfect, was at least familiar and straightforward. Appreciation of the Mercator projection called for computational savvy, and its effective use required reliable methods for taking bearings and determining position. What's the point of precisely plotted rhumb lines when magnetic declination was unpredictable and longitude estimated using astronomical tables might be off by several degrees? Because all the prerequisites for reliable "Mercator sailing"—most notably, the sextant (for measuring precise angles), the marine chronometer, the nautical almanac, and charts of magnetic declination—were not in place until the late eighteenth century, Wright's tables were ahead of their time by nearly two centuries.

Mariners were not the only doubters. A bitter exchange between mathematics teacher Thomas Haselden and three London entrepreneurs suggests that not all British scientists were quick to adopt Mercator's map as the one true chart. The dispute followed a July 1719 advertisement in a London daily newspaper, the *Post Boy*:

> For the Improvement of Navigation, &c.
>
> THERE is now going forward a compleat Sett of Sea-Charts, including all the known Navigable Parts of the World, and representing them according to the Globe itself, without the gross Errors of the Plain Chart, or the puzzling Difficulties, Absurdities, false Views and Deficiencies which the Authors hereof have (now) proved the Mercator's Projection liable to, and rendering Great Circle-Sailing, as Easy and Practicable as Sailing by the Plain Chart; and for the Satisfaction of any Gentlemen, &c. of the Truth hereof, there are several Specimens of this Work already Printed, Approved, and Recommended by Dr. Edmond Halley, and Captain John Merry, which Specimens are to be seen at any of the Authors, viz. John Harris in Bullhead-Court, Newgate Street; John Senex at the Globe in Salisbury Street, and Henry Willson at the Sieve in the Little Minories.

In a short 1722 book Haselden denounced the "authors" for touting a "Globular Chart" with curved meridians and fraudulently securing a "recommendation certificate" from "the Celebrated Dr. Edmund Halley," whom Haselden addressed in his preface:

> [S]ince 'tis very reasonable to suppose, that a Gentlemen of your Learning is taken up with the Contemplation of the Sublimer Parts of Knowledge, and may therefore be a Stranger to what passes among the lower Class of Mankind, I think it the Duty of every Man, who has a real Concern for the Truth, when he hears so great a Character as yours, prostituted to so vile a Purpose, as that of imposing on the Publick, to let you know it. 'Tis upon this Principle . . . that I presume to set before you some of the Articles that have been made use of to usher, if possible, the New Performance (as 'tis call'd) into the World.

Demonstrating that map projection could be a controversial topic three centuries ago, Haselden made the debate both Manichean and personal in defending "the Mercator's-Chart ... which has stood the Test of many Years, and is now generally receiv'd as the best Way of representing the Surface of the Terraqueous Globe, by all who know the Excellency thereof: [which] the Authors of this Globular Performance, like crafty Politicians, who know the Necessity of getting rid of a Formidible Enemy, before they can secure themselves, have represented not only as Puzling and Difficult, but False." One author's treachery was all the more vile for his having endorsed the Mercator map several years earlier: "I cannot but think it would have been much more for the Reputation of Mr. Henry Wilson [*sic*], the pretended Author of this Globular Chart, if he had continued to recommend (as I can shew under his own Hand, he not long since did) the Mercator's as the ONLY CHART, and had not in so prevaricating a Manner endeavour'd to set both the one and the other in a false Light; for by so doing, he had acted the honester Part, and avoided the just Censure of the knowing World."

To bolster his argument, Haselden described the use of the Mercator chart for fourteen typical navigation tasks. The only concession to his opponents' "many groundless Objections" was an admission that accurately measuring and laying off distances with dividers could be troublesome. As modern textbooks on navigation demonstrate, estimating the length of a diagonal course on a Mercator chart is surprisingly simple if the course is no longer than 1,200 miles and does not extend into polar areas, where scale varies enormously. Merely extend the compass between the two end points, as shown in figure 6.1, and transfer the measurement to the scale of latitude graduated in degrees and minutes along the left or right edge of the chart. Distance is easily estimated because a minute of latitude covers roughly one nautical mile. But because north–south scale varies with latitude, it's important to align the dividers vertically with the middle of the course. For a longer course plotted on a small-scale chart, it's wise to divide the route into sections, estimate distance separately for each, and sum the results.

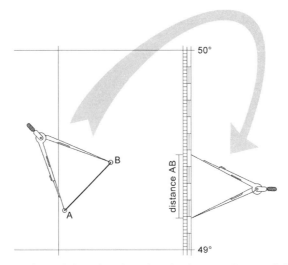

Figure 6.1 Dividers and a latitude scale graduated in degrees and minutes help sailors estimate distance along a diagonal course on a Mercator map.

Navigation handbooks compare Mercator sailing, based on an exact representation of bearings and rhumb lines, with great-circle sailing, which affords a minimum-distance route across a spherical earth. Before radio beacons and electronic navigation simplified great-circle sailing, navigators typically determined a number of intermediate points along a great-circle route, transferred them to a Mercator chart, and sailed the course as a chain of constant-bearing segments. A gnomonic projection, on which great circles are straight lines and vice versa, simplifies the otherwise tedious mathematics of finding intermediate points along a great-circle route. In use as early as the sixth century BC, the gnomonic perspective involves lines of projection radiating from the center of the globe and a tangent plane, which may be positioned anywhere (fig. 6.2, left). Point of tangency is important because scale increases dramatically with distance from the map's center, and a single projection cannot cover a full hemisphere. And as Tissot's indicatrix demonstrates for a map centered in the mid-

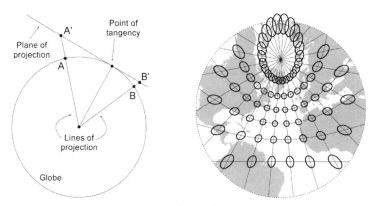

Figure 6.2 In the gnomonic perspective, lines radiating from the globe's center project features onto a tangent plane. Tissot's indicatrix describes areal and angular distortion on a gnomonic projection (right) centered at 45° N, 37° W, between New York and Southampton.

Atlantic (fig. 6.2, right), the gnomonic perspective distorts angles as well as area.

Before electronic navigation, sailing was highly interactive. Because of winds, currents, and intervening obstacles, a ship rarely traveled a perfectly straight course. The navigator had to check his position frequently to confirm that he was on course, track his progress, and estimate time of arrival. Skill in taking measurements, reading tables, and manipulating numbers was essential. Equally important were reliable data on magnetic declination, winds and currents, depth of water, submerged hazards, lighthouses, and other coastal landmarks. For much of this information, sailors relied on hydrographic charts, typically based on a Mercator projection, which provided a convenient framework for data derived from compass readings and carefully measured angles. Edmund Halley, who used Mercator's framework for his 1701 map of magnetic declination in the Atlantic Ocean (fig. 1.6), was an early adopter. Conformality (that is, the absence of angular distortion) was especially helpful to European naval hydrographers, who began to map coastal features using triangulation techniques pioneered by eighteenth-century land surveyors. In the late eighteenth century, when the skills and needs of map users converged with the skills and needs of mapmakers, the Mercator map became the gold standard of marine cartography.

By the mid-nineteenth century, the Mercator projection was so well established that neither Matthew Fontaine Maury (1806–73) nor his biographers considered it worth mentioning. An American naval official hailed as the father of oceanography, Maury used the Mercator projection for maps in his seminal textbook *The Physical Geography of the Sea,* published in 1855, as well as for a set of navigation charts widely credited with making ocean sailing faster and safer. A strong interest in navigation technology and astronomy led to his appointment in 1842 as superintendent of the navy's Depot of Charts and Instruments (later the Naval Observatory), where he discovered a collection of old ships' logs, with daily records of wind and current directions. Curious about world patterns and seasonal effects, Maury summarized the data on "Track Charts" for the Atlantic, Indian, and Pacific oceans, published in early 1848.

Merchant seamen were reluctant to use the charts until a Captain Jackson, sailing out of Baltimore, followed Maury's recommended route to Rio de Janeiro and cut seventeen days off a round trip that normally took fifty-five days. Word of Jackson's voyage spread rapidly, and enthusiastic support among ship owners led to the regular publication of "Wind and Current Charts" in six separate series: Pilot Charts, Storm and Rain Charts, Thermal Charts, Track Charts, Trade-Wind Charts, and Whale Charts—all on a Mercator grid. Eager to make his maps more reliable, Maury struck a deal with merchant captains: Turn over your systematic notes on ocean currents, winds, barometric pressure, air and water temperature, and (of course) position, and I'll give you a free set of our most recent charts. Many complied, and those who did not happily purchased new charts. (Naval captains, who had little choice, were equally eager.) Between 1848 and 1861, when Maury resigned to join the Confederate Navy, the Depot issued two hundred thousand Wind and Current Charts. The project also yielded insightful illustrations (fig. 6.3) for his influential text on oceanography.

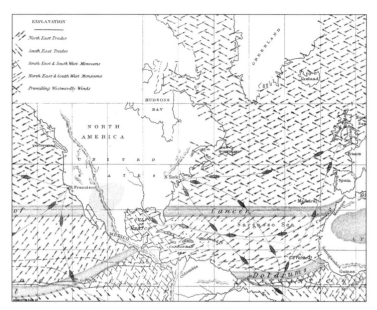

Figure 6.3 Excerpt from Maury's 1855 world map of "Winds and Routes." From Maury, *Physical Geography of the Sea,* pl. 5.

Earlier in his career Maury interacted briefly with another pioneer of American hydrography, Ferdinand Hassler (1770–1843), the Swiss-born mathematician-geodesist hired in 1832 to reorganize the Survey of the Coast (later the U.S. Coast Survey). In 1839, Maury was the most junior lieutenant in the navy. Bored with his assignment, an examination of southern harbors, he wrote Hassler twice, asking to lead a triangulation party, but got nowhere. It's unlikely they talked much when Maury came to Washington in 1842 as the navy's chief hydrographer. Hassler (fig. 6.4) was a feisty fellow, focused on his work and notoriously difficult to get along with. He was not the least afraid of a Congress eager for results and worried about cost, and on one occasion he berated a delegation sent to inspect his shop: "You come to 'spect my vork, eh? . . . You knows notting at all 'bout my vork. How can you 'spect my vork, ven you knows notting? Get out of here; you in my way. Congress be von big vool to send you to 'spect my vork. I

Figure 6.4 Ferdinand Rudolph Hassler. From "Coast Survey," 506.

'ave no time to vaste vith such as knows notting vat I am 'bout. Go back to Congress and tell dem vat I say." No federal official I know would be as fearless or foolhardy.

Hassler's disdain for Washington lawmakers was partly rooted in his earlier service as the survey's first superintendent. Officially appointed in June 1816, he was abruptly terminated in April 1818, when a mercurial Congress transferred the work to the navy. Reduced to odd jobs as a teacher and land surveyor, he continued to refine his strategy for coastal surveys, which he published in 1825 in the *Transactions of the American Philosophical Society.* In 1832, when Congress revived the Survey of the Coast as part of the Treasury Department, he was the obvious choice as its superintendent—a post he held simultaneously with an earlier appointment, in 1830, to oversee the Treasury Department's Office of Weights and Measures.

Although both the navy and the Coast Survey made maps for navigators, Hassler's effort was confined to large-scale, comparatively detailed charts based on original surveys of shorelines and coastal waters. As the nation's measurement guru and the author of a textbook on analytical trigonometry, Hassler resented congressional busybodies who thought he could save time and money by estimating longitude with a chronometer, like a navigator at sea. His experience in Switzerland, as a geodetic engineer, had taught him the importance of exact measurements and a carefully designed triangulation network. To overcome the dense vegetation of salt marshes and coastal thickets, his field parties used precise theodolites mounted on four-foot-high wooden platforms to measure angles between tall poles several miles away.

Equally important was a map projection that would minimize distortion, particularly the distortion of distance. All flat maps stretch or compress some distances—there is no other way to flatten the earth—but distortion is generally low near a standard line, where the globe touches or intersects the projection's "developable surface" (plane, cone, or cylinder). Hassler was especially impressed with the tangent conic projection, which, by definition, touches the globe along a single standard parallel. A thin belt of low distortion straddles the standard parallel, usually positioned near the center of the mapped region. Why not extend this concept, he reasoned, with a map based on many belts of low distortion—better yet, an infinite number of belts, produced mathematically by an infinite number of cones tangent along an infinite number of standard parallels. His solution was the polyconic projection, sometimes called the American polyconic projection or Hassler's polyconic projection.

If this notion seems farfetched, consider carefully the three cones in cross section on the left side of figure 6.5. Each cone defines a conic projection with its own band of low distortion. As shown in the right side of figure 6.5, the bands can be configured to divide the northern hemisphere into the three zones of relatively low distortion. Although the bands don't fit together perfectly—noticeable gaps intervene—they align conveniently along a central meridian. Doubling the number of cones makes the belts narrower and the gaps thinner. Keep doubling, again and again, until microscopically thin gaps separate an indefinitely large number of infinitesimally narrow belts aligned along

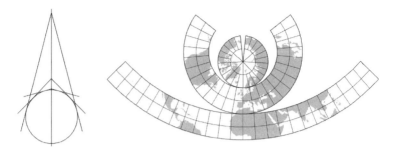

Figure 6.5 A trio of cones tangent at 15° N, 45° N, and 75° N (left) anchor low-distortion conic projections covering thirty-degree belts aligned to a central meridian at 0° (right).

the central meridian. Mathematically, the result is a single projection on which all parallels have true scale.

Is scale really the same along all parallels? Yes, but only in the east–west direction. As Tissot's indicatrix illustrates in figure 6.6, the gaps don't fully disappear. The parallels still diverge as distance from the central meridian increases, but because gaps are not allowed, the map compensates with north–south stretching along the meridians. This stretching is negligible near the central meridian and barely noticeable on a small-scale continental map, except in polar areas, where longitude varies considerably. Circles stretched into ellipses demonstrate that the polyconic projection is not conformal, and the obviously enlarged ellipses on the map's periphery reflect areal distortion as well. But with Hassler's plan, these effects were negligible because each large-scale coastal chart had its own central meridian.

As a local framework for plotting survey measurements, soundings, and topographic details, the polyconic projection was nearly ideal. With east–west scale constant and the central meridian nearby, individual large-scale map sheets had no appreciable variation in scale, area, angles, or direction. For a course beginning and ending on the chart, a straight line represented both rhumb line and great circle, which were too short to betray any measurable departure. Trouble arose when a course extended across two or more charts, or when a mapmaker tried to compile a smaller-scale map covering a much larger

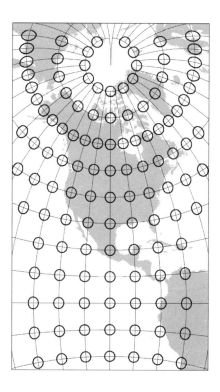

Figure 6.6 Tissot's indicatrix describes areal and angular distortion for a polyconic projection centered on 100° W.

area. Because each chart had its own central meridian, toward which its other meridians converged ever so faintly, charts of adjoining areas immediately to the east or west would not line up. Mapmakers could overcome this difficulty by plotting a single dense grid for the new map and painstakingly transferring features, but navigators much preferred the Mercator chart's standard worldwide grid, anchored at the equator so that adjoining sectional charts at the same scale aligned perfectly. Hassler's polyconic projection was similarly awkward for small-scale sailing charts showing longer courses on a single map sheet: its straight lines were not rhumb lines, its angles were noticeably distorted, and its curvilinear grid thwarted the straightforward reading of latitude and longitude, a simple task with a pair of dividers and the graduated scales along the edges of the Mercator grid.

Although historians attribute the polyconic projection to Ferdinand Hassler, its prominence in American cartography is largely the result of his successors, who not only insisted on a polyconic base for all coastal surveys but also published extensive tables with which other mapmakers could easily lay out a polyconic framework. In his 189-page plan for a systematic coastal survey, published in 1825 in the *Transactions of the American Philosophical Society,* Hassler vaguely alluded to the polyconic projection in the last paragraph: "This distribution of the projection, in an assemblage of sections of surfaces of successive cones, tangents [*sic*] to or cutting a regular succession of parallels, and upon regularly changing central meridians, appeared to me the only one applicable to the coast of the United States." Few charts had been published at the time of Hassler's death in 1843, from a fall and severe exposure while trying to save his instruments during a hailstorm. The Coast Survey's early charts used a simple rectangular projection, no doubt approved by Hassler, and the ostensibly conic framework of charts published in 1844, under his successor, might well be based on tables for the somewhat similar "pseudo-conic" equal-area projection featured in a 1752 maritime atlas by French cartographer Rigobert Bonne (1727–95). Although the Swiss surveyor apparently conceived the polyconic projection around 1820, its widespread use awaited the Coast Survey's publication of a detailed description in 1853 and projection tables in 1856.

Readily available projection tables partly explain the adoption of the polyconic projection by the U.S. Geological Survey, established in 1879. Faced with the enormous challenge of developing reliable base maps for a vast territory only the coastal fringes of which had been systematically surveyed and mapped, USGS topographers could not resist the momentum of more than a quarter-century of precise coastal mapping on a polyconic framework.

Although adequate for piloting harbors and coastal waterways, America's polyconic nautical charts were an annoyance to mariners, who appreciated their accurate shorelines and soundings but preferred a coastal map more geometrically compatible with the chart they used at sea. In 1910, after years of lobbying by the navy, the renamed Coast and Geodetic Survey initiated a program of chart recon-struction. Even so, the Survey's annual report for 1915 indicates that conversion was not equally urgent for all charts: "There is no practical difference except in high latitudes between the Mercator projection and the Polyconic projection, in so far as charts on a scale of 1:80,000 or larger are concerned, but the differences between the projections is appreciable for the smaller scales and is an objectionable feature of the old series of chart." Five years later, when less than half the charts requiring reconstruction had been converted to a Mercator frame-work, a stronger sense of embarrassment reinforced the annual ap-peal for a bigger budget: "Some of our charts . . . are so antiquated as to be of questionable value. They were constructed many years ago on projections which have long since been discredited for naviga-tional use . . . they are on the polyconic instead of the Mercator pro-jection." By 1930 conversion was essentially complete, except for Great Lakes charts, some of which have yet to be converted. Paradox-ically, the Geological Survey did not abandon the polyconic projec-tion until the early 1950s, and coastal hydrographers at NOAA (the National Oceanic and Atmospheric Administration, which was formed in 1970 by combining the Coast and Geodetic Survey, the Weather Bureau, the Bureau of Commercial Fisheries, and several related agen-cies) continued to plot raw survey data on polyconic maps until sev-eral years ago, when digital measurement technology made this inter-mediate step unnecessary by delivering latitude-longitude coordi-nates readily converted to a Mercator framework (or to any other pro-jection, for that matter).

Although the polyconic map was discredited as a navigational tool, cartographic officials at the Coast and Geodetic Survey remained com-mitted to the conic perspective, which is well suited to a mid-latitude region with a pronounced east–west elongation like the conterminous United States. In 1920 they developed a single-sheet national outline map at a scale of 1:5,000,000 using the Lambert conformal projection. With standard parallels at 33° and 45° N, their new base map combined a minimal distortion of distance with a true depiction of angles and in-finitesimally small shapes—ideal properties for the national series of

aeronautical charts that the Coast and Geodetic Survey initiated in
1930 and completed in 1937. Unlike the obsolete polyconic nautical
charts, the ninety-two "sectional airway maps" (fig. 6.7) abutted neatly
along their east and west margins. Scale was not constant—it never is
on a flat map—but commercial pilots considered these deviations far
less troublesome than the corresponding distance variations on a
Mercator projection. At a scale of 1:500,000, the sectional maps covered sufficient territory for convenient flight planning and were sufficiently detailed for "contact piloting" based on major roads, rivers, and
other visible landmarks. Pilots could cut them up and assemble their
own "strip charts," a standard format for aeronautical charts in the
1920s.

Selection of Lambert's conformal conic projection for the sectional airway maps fueled a debate over the relative merits of the Lambert and Mercator projections for aviation cartography. Captain
George Bryan, head of the navy's Hydrographic Office during World

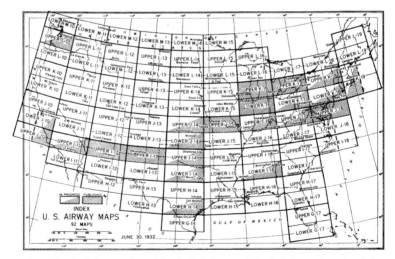

Figure 6.7 An index map (1932) for the U.S. Coast and Geodetic Survey's sectional airways maps. Shading symbols that highlight published charts and work in progress reveal an initial focus on the country's more populous regions. From Ross, "United
States Sectional Airway Maps," 274.

War II, was an unflinching supporter of the Mercator framework, which the navy had consistently favored for charts supporting navigation, whether on water or in the air. Distance measurement was a red herring, he argued, because experts know how to measure distances on a Mercator map, and amateurs can quickly master the graduated bar scale printed on many Mercator charts. Although Lambert and Mercator frameworks are equally efficient for contact piloting, the latter is superior for reading angles, plotting positions, planning courses, and referencing heavenly bodies as landmarks in celestial navigation.

According to Bryan, the Lambert framework's single apparent advantage involved radio bearings, which follow great circles, not rhumb lines. On large-scale maps neither projection needs a correction because straight lines approximate great circles. Medium-scale Lambert projections are also immune because at scales of around 1:1,000,000 the difference between a straight line and a great circle is barely noticeable. But small-scale Lambert charts require a cumbersome correction, much more complex than the corresponding adjustment for a small- or medium-scale Mercator chart. Furthermore, the Lambert projection's medium-scale advantage is largely spurious because in radio navigation the pilot is following a signal, not a map. Better to use one map—a Mercator map—for plotting all navigation data.

Bryan cited endorsements of the Mercator by the Royal Air Force, which considered it the only suitable projection for aeronautical charts, and the International Aeronautical Conference, which in 1919 had approved it as the standard projection for route maps and general aviation maps. Neither recommendation satisfied the Coast and Geodetic Survey and the air force, which collaborated on the 1:1,000,000-scale World Aeronautical Chart (WAC), published on a Lambert conformal conic framework with two standard parallels strategically positioned for low distortion across each sheet. The air force's concern for radio navigation eclipsed the navy's traditional reverence for the Mercator map.

Buy an aeronautical chart these days, and you'll most likely discover its projection is a locally secant Lambert conformal conic, which readily satisfies the International Civil Aviation Organization's flexible requirement for "a conformal projection on which a straight line approximates a great circle." But for areas north of 80° N or south of 80° S, the projection will probably be the polar stereographic, an appropriate substitute for both a locally secant Lambert chart, which is highly similar, and a Mercator chart, which is virtually useless near the poles. A polar gnomonic projection might seem the logical choice, but charting experts consider the polar stereographic's correct angles more useful than the gnomonic's perfectly straight great circles.

Although most cartographic genres have confronted radical technological change, few have had to adjust as rapidly and frequently as aeronautical charting. Early in the last century, when slow, low-altitude flying was the norm, pilots were content with any topographic map showing features readily visible from the air. Increased airspeeds, higher flying altitudes, and better navigation instruments called for more specialized charts focusing on airports, key landmarks, radio beacons, vertical obstructions, and restricted areas. Jet aircraft able to leap several thousand miles in a single flight demanded charts covering greater distances at smaller scales. Automatic piloting, instrument landing, LORAN (*Lo*ng *Ra*nge *Na*vigation), satellite tracking, helicopters, ultralights, gliders, and a host of FAA (Federal Aviation Administration) restrictions added to the complexity and altered the appearance of aeronautical charts. Keeping the charts up to date became far more important than debating the relative merits of similarly suitable conformal projections.

Although weather maps are even more complex and varied than aeronautical charts, meteorologists resolved their search for appropriate map projections more quickly and decisively, through a single international group, the International Meteorological Organization's Commission on Map Projections, which met in Salzburg in 1937. Because meteorologists treat the atmosphere as a phenomenon to be studied, not an obstacle to be traversed, rhumb lines are irrelevant. Far more pertinent are lines describing wind flow and differences in pressure and temperature. Distance is important but angles are more so, especially the angles between isobars and wind arrows and the angles formed where isobars and isotherms intersect meridians and parallels. Accurate depiction of relative direction calls for conformality, which makes the Mercator projection appropriate for tropical areas, close to the tangent parallel at the equator. Similarly, the commission endorsed a conformal polar map based on the polar stereographic projection secant at 60° and a mid-latitude map based on the Lambert conformal conic projection secant at 30° and 60°. For a whole-world map, the commission turned to the Mercator projection, with the caveat that a pair of polar stereographic projections, one for each hemisphere, might be a suitable alternative.

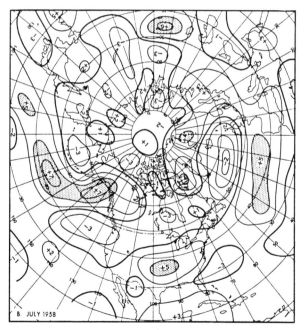

Figure 6.8 U.S. Weather Bureau meteorologists used this hemispherical map on a polar stereographic projection to examine departures from average monthly wind speed (m/sec) for July 1958. Shaded areas reflect a stable jet stream over the Pacific and Atlantic oceans. From Dunn, "Weather and Circulation," 273, fig. 8.

Standardization is important because national weather organizations reap enormous benefits by sharing data with one another, but as the Salzburg report noted, the requirements shouldn't be rigid. For example, researchers exploring upper-level wind velocity might need a polar stereographic map extending well below the Arctic Circle (fig. 6.8). Recognizing the value of flexibility, the commission also endorsed a set of equal-area projections for climatological data, to help viewers compare relative sizes of climatic regions and relate them conveniently to existing equal-area maps of vegetation, soils, and agriculture. The commissioners called for cylindrical, conic, and polar equal-area projections with standard lines identical to their conformal counterparts, but stopped short of endorsing a specific whole-world equal-area projection.

The U.S. Weather Bureau, which had supplied the commission's president, responded promptly, but with no apparent fanfare, by replacing its polyconic map of the United States with a Lambert conformal conic framework. For small-scale newspaper weather charts and similar publications, it's unlikely anyone outside the bureau noticed the change. Visual differences between conic projections offering low distortion can be subtle.

Soldiering On

Read the fine print, my father would say; details matter. Although he had little interest in maps, Dad would have recognized the words and numbers within the white border on U.S. Geological Survey topographic quadrangle maps as minutiae worth examining. Federal mapmakers call this frame the collar. In addition to the sheet name, publication date, map projection, scale bars, and notes about surveys and sources, the collar not only identifies the parallels and meridians that bound the quadrangle but also provides tick marks for three different grid systems. One of the grids is a slight refinement of the earth's spherical framework: two additional parallels and two more meridians partition the quadrangle into nine sections when you join matching tick marks on opposite sides of the map. Because grid lines can interfere with other cartographic symbols, the maps typically show only the grids' intersections with quadrangle boundaries. One set of grid lines running across a map might be tolerable, but not three.

Topographic maps weren't always this way. The 1898 USGS map of Syracuse, New York, that hangs in the half-bathroom off our master bedroom at home shows only the spherical grid, with meridians and parallels portrayed by solid lines, not mere tick marks. It's a 15-minute quadrangle sheet, meaning that the area mapped spans a quarter degree of latitude and a quarter degree of longitude. Interior grid lines are 5 minutes apart, and the projection is polyconic. By contrast, the 7.5-minute Syracuse East quadrangle map published in 1957 contains three grids (fig. 7.1) differentiated with distinctive tick marks. Map users willing to connect corresponding tick marks can plot the spherical grid, the Universal Transverse Mercator (UTM) grid, or the State Plane Coordinate (SPC) grid. The UTM grid, with lines one kilometer apart, is denser than the SPC grid, with a 10,000-foot (3.05 km) separation. The spherical grid, with an interval of 2.5 minutes, is the least dense and least useful. In addition to identifying the various grids, the collar describes the projection as polyconic, which might be wrong— in the late 1950s, when the map was made, the Geological Survey changed projections without updating its map collars.

The story behind map grids and their projections is one of map history's more enigmatic footnotes. Rooted in guns and military engineering, it's a tale of false starts, trickle-down from military mapping into civilian cartography, and the rapid ascent of a previously little-

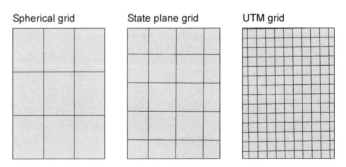

Figure 7.1 The 1957 edition of the Syracuse East, New York, 7.5-minute topographic quadrangle map includes tick marks for three different grid systems. To plot the grid lines shown here, a user must connect corresponding tick marks on opposite sides of the map's collar. Because of differences in the underlying projection, the state plane and UTM grids differ slightly, but noticeably, in their orientation to the spherical grid.

Figure 7.2 A French 75-mm gun and carriage. From U.S. Ordnance Department, *Handbook of Artillery*, 75.

used map projection. It's also a story that links Gerard Mercator with two of the world's all-time mathematical geniuses, Lambert and Gauss, both of whom contributed to the transverse Mercator projection.

If there's an impetus here, it's the secret long-range field gun (fig. 7.2) developed by the French in the 1890s. The mobile, rapid-fire 75-millimeter cannon depended on an innovative hydropneumatic recoil system that absorbed kickback and returned the gun to its original position. With a trained crew, the "French 75" could stay aimed at a target while firing five to thirty rounds per minute. Artillery pieces that matched its five-mile range were not so quickly reloaded because the gun crew had to stay clear of a recoiling cannon, which often had to be re-aimed after a few rounds. Reloading time and firing rate were crucial because the longer the firing period, the more vulnerable a gun was to detection and attack. When hidden, the quick-firing French 75 offered fewer clues to its location.

The increased range of early-twentieth-century artillery triggered a transition from "direct firing" at visible targets to "indirect firing" at positions inferred from telephone reports by spotters stationed along the front. Artillery manuals of the era describe a trial-and-error process called "shooting in." Triangulation based on reports from widely separated observers provided a rough indication of where a shell landed. If it fell short of the target, the crew increased the charge and elevated the barrel. If it landed too far to the right, they aimed the next volley a bit to the left. Indirect fire from a hidden position was the best way to exploit the long reach of a rapid-fire weapon like the French 75.

To further improve their aim, the French enhanced their battle maps with mutually perpendicular sets of evenly spaced parallel lines similar to a sheet of graph paper. The grid gave every feature a pair of rectangular coordinates, an X and a Y, that made it easier to calculate distance and direction from gun to target. Although a mathematician could figure it out using latitude and longitude, rectangular coordinates made gunnery calculations faster and more reliable—simple exercises in analytical geometry and trigonometry. Called "map firing," this more advanced form of indirect fire also gave spotters on the ground or in the air a concise, reliable way to describe enemy positions. Map firing at unobserved targets increased the importance of reliable maps and soldiers who could read them.

Military textbooks describe the roles of grid-based calculations. I found several classic examples in *Military and Naval Maps and Grids, Their Use and Construction,* published in 1942. A typical problem relates three points.

> From an observation post at C (1162.120–2294.210) an enemy supply base B has Y-azimuth 73° 30′ and range 6500 yards. Find the range and Y-azimuth of B from gun A (1161.253–2293.545).

The numbers following point C are rectangular coordinates given in yards, the preferred distance measure on prewar American military maps. By convention, the x-coordinate (1162.120) precedes the y-coordinate (2294.210). Target range and distance calculations are also in yards. The y-azimuth reports directions in degrees clockwise from the more northerly grid direction—as the SPC and UTM grid lines for the Syracuse East map (fig. 7.1) demonstrate, grid systems seldom align perfectly with parallels and meridians.

In another example a moving target underscores the importance of rapid calculations.

You are in command of a field gun whose position is 996.000–
2046.000 and you are at 969.000–2045.600. You observe an enemy
tank at a distance of 8000 yards from you whose Y-azimuth from
your position is 54° 45′. In what direction and at what range will you
have the gun set to fire on the tank?

A further variation invokes the speed of sound, assumed to be 1,100
feet per second.

Your gun A is at 1046.320–2010.460 and gun B is at 1051.390–
2003.510. After seeing the flash as an enemy gun south of AB the
sound of the explosion is heard at A 17 seconds later and at B 21 seconds later. Determine graphically the range and direction for each of
your guns in order that they may fire on the enemy gun.

As this last example suggests, a gridded map or sheet of graph paper
affords a more rapid (but less precise) graphic solution.

Map grids work well over small areas, for which earth curvature is
conveniently ignored. Choose the right projection, and a reliable grid
might extend hundreds of miles in one direction. At some point,
though, it's necessary to define a new grid, anchored nearby on a new
projection. The area within which a grid affords reliable estimates of
distance and direction is called its zone.

The most basic grid system consists of a zone, a map projection,
two sets of grid lines, and a numbering system. Centering the projection within the zone keeps distortion low, while an origin (zero point
for grid coordinates) outside the zone assures that all coordinates are
positive, as in figure 7.3. This "false origin" eliminates the likelihood of
a salvo going awry because a soldier forgot the minus sign. Because
the y-axis points generally northward, the y-coordinates are called
northings. And with the x-axis pointing eastward, the x-coordinates
become eastings. When American military cartographers established
the UTM system in the late 1940s, they adopted the European convention of indicating northings and eastings in meters.

Grid lines through the projection's center typically coincide with a

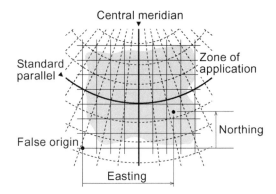

Figure 7.3 The key elements of a rectangular coordinate system.

meridian and a parallel so that here at least the direction "grid north" reflects true north. Away from the origin, earth curvature deflects grid north away from true north and grid east away from true east. Typical of map distortions, this deflection increases with distance from the projection's axes.

For mid-latitude countries like France and the United States, map grids are usually based on either a conic or a transverse cylindrical projection. A conic projection with two standard parallels supports a low-distortion grid with a pronounced east–west elongation. Lambert's conformal conic projection with two standard parallels (fig. 7.4) is especially useful because a cartographer can constrain maximum distance distortion within the zone by carefully adjusting standard parallels and zone boundaries. In the mid-1930s the U.S. Coast and Geodetic Survey used the Lambert conformal conic framework to design east–west trending zones for the State Plane Coordinate system. A key SPC requirement is that a distance calculated using grid coordinates may not differ from the true distance across a round earth by more than one part in ten thousand, the precision of high-quality surveying instruments of that time. A zone too large to satisfy this constraint for all pairs of points within the zone must be split in two. This specification explains why Tennessee has a single SPC zone while Pennsylvania, not as narrow north to south, requires two zones.

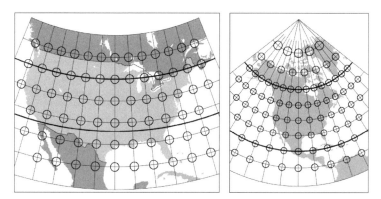

Figure 7.4 Tissot's indicatrix illustrates generally low areal distortion on Lambert conformal conic projections with standard parallels (heavy lines) tailored to the conterminous United States (left) and North America (right). Because of strict limits on scale error, the State Plane Coordinate system employs much smaller zones with markedly closer standard parallels.

For zones with a north–south elongation, the SPC system relies on the transverse Mercator projection, which maps the globe onto a cylinder centered on two opposing meridians, as shown in figure 7.5. With an appropriately positioned central meridian as its standard line, the transverse Mercator affords a low-distortion framework for narrow north–south zones only a few degrees wide. In the same way that Pennsylvania is too wide, north to south, for a single Lambert-based zone, Illinois is too wide, east to west, for a single transverse Mercator zone. Conformal like its equatorially centered parent, a transverse Mercator projection incurs massive areal enlargement when extended more than eighty degrees to the east or west of its central meridian. Lacking the equatorial Mercator's essential asset— straight lines on the transverse Mercator do not represent rhumb lines—it's seldom used (except in books about map projection) for small-scale maps.

Before World War I, French artillery officers relied on independent local grids based on Bonne's projection (a nonconformal polyconic variant) and centered on strongholds from which fixed guns might

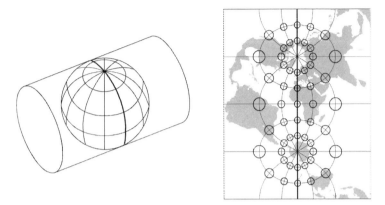

Figure 7.5 The transverse Mercator projection, developed on a cylinder tangent to the globe along two opposing meridians (left), yields a conformal map with low areal distortion for a tall, thin zone surrounding a central meridian (right). As Tissot's indicatrix suggests, distortion increases rapidly toward the map's eastern and western edges, where scale approaches infinity.

conveniently bombard targets in the region. After the angular distortions and awkward discontinuities of the Bonne grids became apparent early in the war, French officials devised a single military grid for the likely battle zone in the northern part of the country. Based on a Lambert conformal conic projection, the integrated "Nord de Guerre" was easily extended eastward into Germany as well as westward toward the English Channel. Impressed with the grid's effectiveness, the French used a slightly modified version in World War II.

Directionally accurate long-range artillery and "map firing" also established a need for military surveyors, who relied on conformal projections in helping gun crews get a fix on true north by tying the gun's position into a precisely measured triangulation network. Magnetic compasses could not be trusted, and triangulating with visible landmarks was difficult when the gun was hidden.

The French military was more cartographically astute than their German foes, who had little appreciation of map grids until they captured a map with a Bonne grid centered on Verdun. Previously content with trial-and-error indirect firing, the Germans adopted the Verdun grid and naively extended it westward one hundred kilometers (62 mi) to Rheims. When this expanded zone proved too large and imprecise, they devised a system of disconnected local grids based on the Cassini projection, a nonconformal framework reliable only within 60 kilometers (37 mi) of its central meridian. In fall 1915 the German military survey called for a unified coordinate system based on the transverse Mercator projection, appropriately conformal, but field commanders and the general staff, who anticipated a quick victory, resisted the change.

With more reliable map grids, the Germans might have made better use of the rail-mounted long-range artillery that bombarded Paris late in the war. The infamous "Paris Gun," also known as "Large Max," weighed 138 tons and had a barrel 112 feet long, a muzzle velocity of more than 5,000 feet per second, and a range of 75 miles. Intended to terrorize French civilians, the massive cannon killed 250 people with 360 shells fired intermittently between March 1918 and August of that year, when it was evacuated to Germany and dismantled.

Little used before World War I, the transverse Mercator projection was one of three conformal projections presented in 1772 by Johann Heinrich Lambert, who illustrated its assets with a small-scale map of the Americas. In 1822, the German mathematician Carl Friedrich Gauss (1777–1855), at work on a highly precise survey of Hannover, presented formulas for a more exact transverse Mercator framework, based on an oblate globe, flattened slightly at the poles and a bit broader at the equator. Historians trace this refinement of the earth's shape to Isaac Newton (1642–1727), who deduced a slight flattening of the earth in a discussion of rotating bodies in his monumental *Principia Mathematica,* published in 1687. Triangulation surveys verified Newton's theory a half-century later, and we now know that the earth's radius is about 1/297 longer at the equator than at the poles. After nineteenth-century geodesists estimated the earth's polar and equatorial radii, Gauss's formulas for the transverse Mercator projection allowed a more reliable rendering of distance on large-scale maps with a north–south elongation. In addition to devising this low-distortion framework for large-scale maps, Gauss introduced the term *conformal.*

Although highly useful, Gauss's refinement was neither original nor unique. Lambert had included ellipsoidal formulas for his conformal conic projection, and the U.S. Coast and Geodetic Survey published ellipsoidal tables for the polyconic projection in 1854. Other projections, including several that preserved relative area, had ellipsoidal formulas, but their properties were irrelevant to artillery calculations, which focus on range (distance) and azimuth (direction).

For military maps a projection's suitability hinges on whether grid-based calculations should favor distance or direction. In general, a polyconic projection centered within the zone yields slightly more accurate estimates of distance while a locally centered conformal projection provides marginally more reliable estimates of direction. But because properly tailored conformal projections do not distort distances as much as polyconic projections distort angles, much is gained and little lost by favoring highly reliable angles.

Jacob Skop, head map designer at the U.S. Army Map Service, compared these trade-offs in a 1951 article in *The Military Engineer.* A conformal projection might over- or underestimate distance to a target ten thousand yards away by a mere two yards—insignificant in comparison to the twenty-two-yard uncertainty of a six-inch gun at that range. By contrast, at this range a polyconic projection might yield a deflection error as large as twelve yards, well above the gun's likely azimuthal error of two or three yards, and more than enough, as Skop noted, to "seriously impair the effectiveness of the weapon." What's more, the miss would be proportionately greater for a target farther away.

In light of French and German experiences during World War I, when American artillery brigades used French maps as well as the French 75, it's puzzling that in 1918 the United States adopted the polyconic projection for a military grid that divided the conterminous states into seven zones centered on meridians eight degrees apart. Although the north–south elongation of the zones was ideal for a transverse Mercator projection, the chief architect of the U.S. Polyconic

Grid was William Bowie (1872–1940), head geodesist at the U.S. Coast and Geodetic Survey, where the polyconic framework was a cornerstone of the Hassler legacy. Bowie was surely aware of the projection's lack of conformality, but his description of the grid ignored directional error altogether and declared the scale error "negligible" relative to paper swelling caused by atmospheric humidity. Although he served as a major in the Army Corps of Engineers between August 1918 and February 1919, Bowie apparently had little opportunity or inclination to seek the advice of artillery experts. Jacob Skop, who labeled the choice of the polyconic projection "unfortunate," observed that "azimuth errors at the meridional limits of the zones were far in excess of permissible tolerances." Fortunately, American troops never had to fight with long-range artillery on their own soil.

In 1947 the army rectified its mistake of 1918 with the Universal Transverse Mercator grid. Affording worldwide coverage of the interpolar area between 84° N and 80° S with sixty north–south zones six degrees of longitude wide, the UTM system relies on a secant case (see page 27) of the transverse Mercator projection. On an unmodified, tangent version of the projection, map scale increases with distance from the central meridian, which is the only line of true scale. To create two lines of zero distortion within the zone, army cartographers reduced the scale along the central meridian to 0.9996 of the map's stated scale. This adjustment distributes distortion more evenly across the zone (fig. 7.6) and lowers the average distance error, which otherwise would almost always be slightly inflated. Directional accuracy is not affected because the projection remains conformal. Six-degree zones hold distance error to less than one part in 2,500 in Europe, where Cold War strategists deemed future battles most likely, thereby minimizing what Army Map Service official John O'Keefe called "the evil effects of zone junctions." Although the Lambert conformal conic projection already had two lines of true scale, a worldwide Lambert system would have required over two hundred zones.

A legacy of the French 75 was an increased appreciation of conformality among cartographers, most notably Arthur Hinks, a leading British authority on maps and surveys. In the preface to his innovative

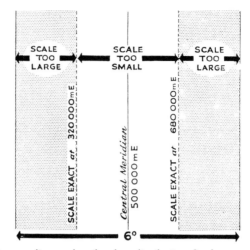

Figure 7.6 An army diagram describes the redistribution of scale errors on the secant transverse Mercator projection used for the UTM grid. On a tangent version, scale would be "too large" everywhere except along the central meridian. As with all UTM coordinates, eastings for the central meridian and the two lines of true scale are in meters. From Robison, "Military Grids," 6.

textbook *Map Projections,* published in 1912, the Cambridge scholar confidently asserted "the property of orthomorphism (conformality), which plays such a large and difficult part in the theory of Map Projections, is not in most cases of any great advantage or importance in actual mapmaking." Nine years later, Hinks introduced his second edition with a retraction: "the requirements of the artillery in modern war have brought into great prominence the advantages of an orthomorphic (conformal) projection for the large scale tactical maps used in stationary warfare; and what I said of orthomorphism in 1912 needs modification."

Similar sentiments underlie the choice of conformal projections for the State Plane Coordinate system, for which grid ticks began appearing on American topographic maps in the late 1930s. Ironically, the U.S. Geological Survey, perhaps seduced by the country's relative isolation, resisted conformality for its large-scale base maps until the

1950s, when it quietly dumped the polyconic projection and began converting its quadrangle maps to the conformal projection, Lambert or transverse Mercator, for the area's SPC zone—in many cases without noting the change in the map collar. Whether the bureaucratic inertia behind the polyconic's persistence explains its even longer nominal longevity is anybody's guess.

On Track

In much the same way that rapid-fire artillery triggered wider interest in Lambert's largely dormant transverse Mercator projection, long-distance flying created a new role for the oblique Mercator projection, centered on a great circle that is neither the equator nor a meridian (fig. 8.1). In addition to establishing a low-distortion corridor along a particular great-circle route, an oblique cylinder provides a more geometrically accurate view of a country, continent, or traverse with a pronounced elongation that is neither north–south nor east–west. When federal cartographers set up a State Plane Coordinate system for Alaska in the early 1960s, the oblique Mercator, which is conformal, offered a conveniently efficient single-zone solution for the new state's Panhandle, over five hundred miles long and inclined diagonally. In the early 1970s, satellite remote sensing precipitated a further, more revolutionary modification, the Space Oblique Mercator projection, which provides a reliable geometric framework for aerial imagery captured from an altitude of several hundred miles by an electronic scanner in a steadily shifting orbit.

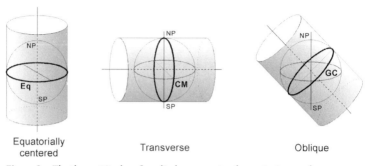

Figure 8.1 The three attitudes of a cylinder can center the projection on the equator (left), a central meridian (center), or an inclined great circle chosen as an axis of low distortion (right).

Perhaps the earliest use of an oblique Mercator projection, or indeed any oblique cylindrical projection, was for maps of Central America and Southeast Asia in a world atlas published by the innovative German mapmaker Ernst Debes (1840–1923) in 1895, a year after American polymath Charles Sanders Peirce (1839–1914) quietly, and apparently independently, sought a patent for his "Skew Mercator" projection. Although the oblique attitude was well suited to Central America's northwest–southeast trend, the elongation is too subdued to confer a clear advantage over more traditional north-at-the-top portrayals.

Oblique customization of the conformal cylindrical framework languished until the 1920s, when pioneering transcontinental flights called for thin maps, often with a decidedly diagonal slant. In March 1921, in what might be the first use of an oblique Mercator projection to describe a long flight, *National Geographic* magazine used an "inclined cylindrical projection" to describe Sir Ross Smith's first-person account of his thirty-seven-leg, eleven-thousand-mile "aëroplane voyage" from London to Adelaide, Australia. Stretching across the upper half of two facing pages, the customized perspective wasted less space than a conventional layout. In celebrating Smith's route with a low-distortion oblique Mercator framework, the magazine's mapmakers rejected the obvious shortcut of cutting a diagonally oriented strip out of a readily available north-up map.

Complementing Smith's 111-page narrative, a shorter article by air power advocate General William (Billy) Mitchell predicted high-altitude transoceanic flights at speeds up to 400 miles per hour. Eighteen years later Pan American World Airways validated Mitchell's vision by inaugurating regularly scheduled service between New York and Southampton, England. Because of the distances involved, the airline favored a great-circle route, which was not only appreciably shorter than a rhumb line but allowed pilots to lock onto radio beacons, which followed great circles. Although World War II intervened, postwar airlines quickly reestablished long-distance service along great-circle routes. To reduce fuel load, most transatlantic flights from North America stopped at Gander, Newfoundland, less than 2,000 miles from Shannon, Ireland.

In early 1947 the U.S. Coast and Geodetic Survey recognized the growing preference for high-altitude great-circle flying and the importance of Gander as a refueling stop by issuing a regional aeronautical chart on an oblique Mercator projection tangent along the great circle linking Chicago and Gander. Published at a scale of 1:2 million and formatted with its standard line running diagonally, as in figure 8.2, the 26 by 54 inch (66 by 137 cm) chart included key East Coast airports from Washington to Boston, which also originated flights through Gander. By holding scale distortion along the coast to less than one percent, the Chicago–Gander axis allowed a single map to serve the Atlantic side of transoceanic operations. And because the projection was conformal and its tangent line close by, great circles were nearly straight lines and radio bearings required no correction.

A quarter-century later aerospace engineers posed a new cartographic challenge: how to generate an appropriate geometric framework for satellite imagery captured by low-altitude earth resources satellites like Landsat-1, launched in 1972. Unlike a telecommunications satellite, which feeds thousands of dish antennas from a fixed position in "geostationary" orbit 22,300 miles (36,000 km) above the

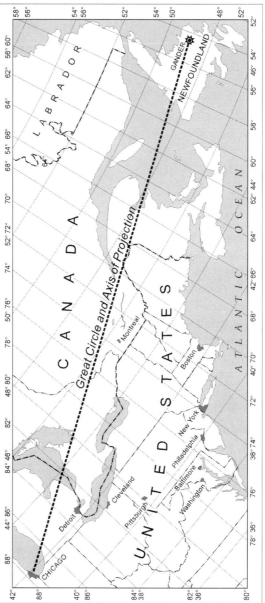

Figure 8.2 The *Military Engineer* for February 1948 included this much-reduced description of the U.S. Coast and Geodetic Survey's Chicago–Gander aeronautical chart based on an oblique Mercator projection. Redrawn from Stanley, "Map Projections for Modern Charting," 57.

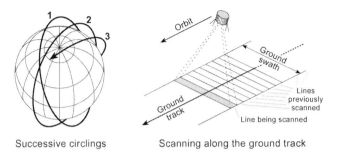

Successive circlings   Scanning along the ground track

Figure 8.3 Schematic drawings of three successive circlings (left) of a low-altitude satellite in a precessing sun-synchronous orbit and the ground image (right) built up by adding scan lines along a ground swath perpendicular to the satellite's ground track.

equator, a mapping satellite circles the globe more than a dozen times a day at an altitude of approximately five hundred miles—much closer to the ground for greater detail—while its imaging system adds row after row to a continuously evolving array-like picture centered on a gently curved ground track that describes the satellite's progress across the surface. Figure 8.3 illustrates this progressive accumulation of a long, thin image for a ground swath winding around the globe.

Aerospace engineers call this circuitous path a sun-synchronous precessing orbit. The orbital plane, tilted about nine degrees away from the earth's axis, moves slowly westward so that the satellite always crosses the equator at the same local sun time, typically in mid-morning. Although synchronized equator crossings minimize variation in sunlight, the scanner can examine only a fraction of the earth's surface in a single day. To insure wider coverage, the next day's ground tracks fall slightly farther west of those imaged the day before so that eventually the satellite scans the entire planet, except for polar areas, at least once before repeating the pattern. For Landsat-1, with an altitude of 570 miles (923 km), an equator crossing every 103 minutes, and a ground swath of 115 miles (185 km wide), the "return interval" was eighteen days. Landsat-7, launched in 1999, follows a generally similar pattern from an altitude of 438 miles (705 km) but repeats the cycle every sixteen days.

In addition to generating a steady stream of land-cover data, continuous scanning introduced the geometric complexity of a moving vantage point. By comparison, conventional air photos, which had revolutionized topographic mapping in the 1930s, were mere snapshots: perspective views of terrain, which photogrammetric technicians could straightforwardly (if not quickly) convert to flat maps. Although Landsat's innovative geometry was demanding, mapping scientists focused on the larger challenge of classifying "multi-band" satellite data with visible and infrared components. With a ground resolution of 79 meters (260 feet), several hundred times grainier than conventional aerial photography, Landsat imagery was far more appropriate for adding detail to existing coverage than for making maps from scratch. By applying standard geometric corrections, savvy users could align meridians and parallels to Landsat images or overlay individual pixels (picture elements) on existing maps. Because image resolution was coarse by cartographic standards, approximate positions estimated with "rubber sheeting" were fully adequate.

Civilian mapping scientists who foresaw saw wider possibilities for remote sensing were already contemplating mapping satellites with high-resolution imaging systems suitable for topographic mapping. Among these visionaries was Alden Colvocoresses, the U.S. Geological Survey geodesist who patented MapSat in 1979 and assigned rights to his employer. Well aware of the finer resolution of top-secret intelligence satellites—by 1976 electronic space imagery with a resolution of 6 inches (15 cm) was helping the CIA monitor missile development in Russia and China—Colvocoresses proposed assigning pixels explicit positions on an inclined cylindrical projection that recognized the combined motions of the rotating earth, the orbiting scanner, and the precessing orbit (fig. 8.4). In a 1974 *Photogrammetric Engineering* article laying out specifications for a conformal framework he dubbed the Space Oblique Mercator projection, Colvocoresses conceded that neither he nor his colleagues had worked out the formulas. If cartographers could develop a rigorous mathematical relationship between the precessing cylinder and the ellipsoid, he ar-

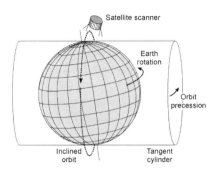

Figure 8.4 Unique elements of the
Space Oblique Mercator projection.
Redrawn from Colvocoresses,
"Space Oblique Mercator," 921, fig. 1.

gued, the "era of automated mapping, based on earth-sensing space systems, is not far off."

Two years later, in Columbus, Ohio, at an international symposium on "The Changing World of Geodetic Science," Colvocoresses repeated this appeal. His audience included John Snyder (fig. 8.5), a chemical engineer on vacation from his job with a pharmaceutical manufacturer in northern New Jersey. Author of a carefully researched book on his state's cartographic history, Snyder was an avowed map enthusiast and self-taught expert on map projection. Intrigued by the challenge, he set to work on a solution that consumed much of his free time over the next few months. Drawing on spherical trigonometry, analytical geometry, and calculus as well as his skill with a programmable calculator, he first solved the moving vantage point for a sphere and then, with a month's more work, extended his solution to the ellipsoid—an essential refinement for precise space-based mapping. Satisfied with the results, he sent Colvocoresses a package of eighty-two equations and sample calculations. After scrutinizing Snyder's mathematics and converting his formulas to a mainframe computer, USGS geodesists confirmed that the New Jersey novice had outshone a team of mathematical consultants.

Equally surprising, the answer was free. The Geological Survey, which had spent $22,000 on consultants' fees and was apparently willing to invest further in its quest for a solution, expressed its gratitude

Figure 8.5 John Parr Snyder (1926–97). Courtesy of the U.S. Library of Congress.

in 1978 with a John Wesley Powell Award for Citizen Achievement. In noting that Snyder had "provided the sought-after link by which earth surface data obtained from orbiting spacecraft can now be transformed to any of the common map projections," the citation referred to the projection's inverse formulas, with which analysts could recover spherical coordinates (latitude and longitude) from a satellite image and plot features on a conventional Lambert or transverse Mercator framework.

Aware of Snyder's unique talent and energy, the USGS eagerly hired him as a consultant to advise on a range of issues involving projections, grid systems, and map accuracy. Recognizing a rare opportunity, Snyder worked out a part-time arrangement with his employer in New Jersey. Two years later, after the Geological Survey offered him a staff scientist position, he began a new career, in his mid-fifties, as a full-time mapmaker. John Snyder quickly became a leading figure in American cartography. In addition to numerous USGS publications on map projections, he wrote the definitive history of the subject, coauthored several textbooks, and served as president of the American Cartographic Association.

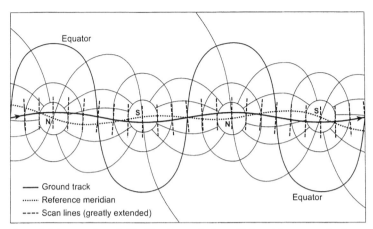

Figure 8.6 A one-and-one-half orbit snippet of a Space Oblique Mercator projection with a thirty-degree graticule illustrates the progressive displacement of the ground track. Compare the ground track with an arbitrary reference meridian, and note that between the satellite's first and second southward sweeps the ground track moves from the east to the west of this reference line. Scan lines, extended for clarity, exaggerate the width of the ground swath. Redrawn from Snyder, *Map Projections Used by the U.S. Geological Survey*, 196.

Interviewed by the *New York Times* shortly after receiving the Powell award, Snyder was characteristically modest. "I figured I was the last person who would be able to solve the problem," he told the reporter. "But I'd spent a lot of time with my calculator, making an intensive study of map projection formulas, and I'd run out of simple things to do." He readily acknowledged the cleverness of Colvocoresses, who framed the problem, and the effort of University of Virginia professor John Junkins, who also sent in a solution. Although the professor's formulas accommodated an eccentric, noncircular orbit used for some spy satellites, his solution was more complicated, less precise, and clearly not conformal. The Space Oblique Mercator projection, Snyder noted, required some computational shortcuts, and his own solution, while more accurate, was not completely error free. Even so, he told the *Times*, "scale factors are within a few millionths of correct values, and are thus well within mapping accuracy."

In a 1978 article for *Photogrammetric Engineering and Remote Sensing*—renamed to recognize the revolutionary implications of space imagery—Snyder offered a general description of the projection's geometry and development. An example (fig. 8.6) spanning one and a half orbits illustrates the progressive westward migration of the ground track, noticeably displaced for the second orbit. With the static oblique Mercator projection as a starting point, his derivation introduces the complexity of the precessing orbit and the need to consider the rotating earth's effect on the scan lines, which are not fully perpendicular to the gently curving ground track. Although scale errors remain less than one part in ten thousand within the ground swath—the cartographic standard adopted in the 1930s for the State Plane Coordinate system—distortion increases with distance from the ground track. Because many of the formulas are amenable only to iterative, trial-and-error solutions, whereby the computer converges bit by bit on an acceptable result, the projection is "not . . . exactly conformal." Hardly a problem, though, because the added precision of an exact solution would be useless.

Like most customized map projections devised in the late twentieth century, the Space Oblique Mercator projection is partly a consequence of electronic computing. Snyder's acknowledgments, at the end of his article, underscore the equally important role of camaraderie and self-confidence. After thanking Colvocoresses, Junkins, and the three USGS programmers who validated his formulas, he confessed he "would not have undertaken this derivation . . . without the initial encouragement of Prof. Waldo R. Tobler and the technology available in hand calculators."

Wall Maps and Worldviews

Persistent misuse of the equatorial Mercator projection, especially for world maps having nothing to do with navigation, taunts cartographically savvy geographers. Bad enough when we spot one decorating a nightly newscast or hanging in a small shop struggling to look cosmopolitan. Worse yet when we find a blatant example near home, as I did in the mid-1990s, during parents' night at my daughter's middle school. But there it was, hanging from a spring roller in Jo's social studies classroom: a huge wall map with bright blue oceans, richly colored countries, and a Greenland bigger than China. The teacher, who saw nothing amiss, shrugged when I asked who ordered it and from where. We rely on educators' supplies catalogs, she replied, and since the map came from a reputable dealer, it must be right. *Right?*

Tempted to write the principal or school board, I heeded Jo's advice that parents ought not mimic the self-appointed morality police who condemn public libraries for stocking Judy Blume or J. D. Salinger. Besides, she argued, the teachers rarely refer to the map, even in "global studies," the New York State Education Department's naive

strategy for reintroducing geography without requiring teacher certi-
fication in the subject. Another ax to grind.

I don't know whether the map is still there, subtly reinforcing mis-
perceptions of an India smaller than Scandinavia, but Mercator reference maps are alive and well in teachers' supplies catalogs, online as
well as in print. (The prodigious Google search engine turned up several Web sites offering Mercator wall maps, but only one, oddly, identified the projection by name.) I don't mean to imply that all or most
schools are misrepresenting the world—broad availability, after all,
does not demonstrate widespread use—but ill-informed retailers front-
ing for vendors who should know better can easily dupe unaware buyers. Although most wall-map catalogs offer world maps on other projections, the equatorial Mercator world map is pervasively entrenched
in the wall-map trade (fig. 9.1).

We can trace the roots of this persistence to the eighteenth century, when navigators and explorers belatedly adopted the Mercator
worldview. Revered by mariners and readily at hand, the projection
provided a convenient if inappropriate framework for land-focused
maps, on which rhumb lines are pointless. A prominent example is
Henry Popple's 1733 map of British possessions in North America,
printed in twenty sheets that could, as an option, be pasted on linen,
equipped with rollers, and hung as a wall map, eight feet wide and
eight and a half feet tall. Sold by subscription—two Guineas down,
two more on delivery—it was advertised in London's *Daily Post* as
"laid down according to Mercator's Projection, to be engraved by the
best Masters, and printed upon the best Paper." Although the original
version sold poorly, Popple's successors cut the price in 1739 and enjoyed comparatively brisk sales during a three-year war with Spain
over what is now Florida and Georgia. Acclaimed by map historians as
one of the two most significant large-scale maps of colonial North
America, Popple's map was not only copied more than a dozen times
but also served as a source for smaller-scale maps in atlases and textbooks.

With name recognition and a distinctive appearance, the equatorial
Mercator projection became the standard world map for nineteenth-

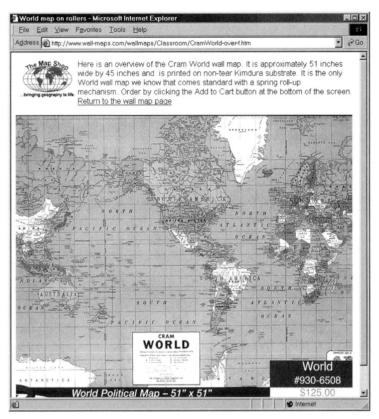

Figure 9.1 This Web advertisement for a Mercator wall map does not identify the projection.

century atlases and wall maps. Not immediately, though: some atlas publishers preferred a pair of globular views, with separate spotlights on the eastern and western hemispheres, while others complemented the equatorial Mercator perspective with these twin hemispheres. By contrast, wall-map publishers readily embraced the Mercator map's rectangular format, which conveniently matched the straight lines and right angles of the typical wall space. Schoolbooks and classroom atlases also promoted the Mercator worldview. Jedidiah Morse (1761–

1826), a prolific author of popular school geographies, used only one world map, on a Mercator projection, in his *Compendium and Complete System of Modern Geography,* published in Boston in 1814. His son, Sidney Edwards Morse (1794–1871), who took over the family business, inherited Jedidiah's affinity for the Mercator map. Sidney's development of wax engraving, an efficient method for enriching maps with small but legible place and feature names, helped promote Morse and Mercator as geographic brand names.

An emerging preference for the Mercator worldview is apparent in John Snyder's cursory survey of thirteen world atlases published in the United States, Britain, France, or Germany between 1820 and 1897. To explore the relative popularity of specific projections, Snyder tabulated the projections used for whole-world maps and for separate maps of eleven major world regions. Although all thirteen atlases employed a variety of cartographic frameworks, most favored a specific projection for each region. For example, ten of the atlases cast their map of Africa on a sinusoidal grid, while nine framed their twin maps of the eastern and western hemispheres on a globular projection. Amid this mild diversity, all of the nine atlases with a second whole-world map relied on the Mercator grid. Moreover, of the eleven atlases with a separate map of Oceania, nine had a Mercator framework, an inappropriate choice despite the region's broad expanse of water. Because few buyers would ever plot rhumb lines or estimate bearings with a world atlas, this distorted picture of the Pacific was a disservice to anyone interested in the relative separations of ports and islands.

A similar tabulation, based on sixteen world atlases published in the United States, Britain, Germany, or Russia between 1916 and 1990, documents the decline of the Mercator projection in the latter half of the twentieth century. Of the six atlases with a whole-world Mercator projection, the most recent was published in 1966. And among these six, only the three earliest, issued between 1916 and 1941, used only a Mercator framework for their world maps. By contrast, the three more recent atlases, in addition to a casting a world map on the Mercator grid, included world maps on at least one other projection, such as the polar azimuthal equidistant or the sinusoidal. At the regional level, the Mercator projection framed maps of Oceania as late as 1941 and survived into the twenty-first century in National Geographic atlases that use it for separate treatments of the Atlantic, Indian, and Pacific Oceans.

Two additional surveys attest to the Mercator's decline. Arthur Hinks, in his 1912 textbook on map projection, examined ten general reference atlases published in Britain, France, or Germany between 1894 and 1912. In addition to noting that two of the six atlases with maps of Oceania employed the Mercator projection, he observed that the "Mercator, Globular, and Mollweide are used by nearly all of the atlases." A half-century later Syracuse University master's student Frank Wong examined thirty-eight world atlases published in the United States between 1940 and 1960. Although the Mercator projection dominated whole-world atlas maps before 1940, a spate of new projections precipitated its decline, but not demise, during the 1940s. Distribution maps focusing on population and economic activity were the first to be revised. Some publishers substituted an equal-area map while others compromised with projections that were neither equal area nor conformal. Political maps focusing on boundaries and place names held out a bit longer. So strong was the Mercator worldview that initially, during the early 1940s, most atlas makers cautiously added another world map to supplement, rather than replace, the familiar framework. Even so, the transition was nearly complete by 1951, when all American atlas publishers but the C. S. Hammond Company had substituted another projection for their world political maps.

There's little evidence that atlas publishers, in largely abandoning the Mercator projection during the 1940s, were finally heeding the admonitions of scholarly critics. As early as 1912 the influential cartographic educator Arthur Hinks had railed against "the great distortion in the north and south [that] makes Mercator's projection altogether unsuitable for a land map." Nine years later U.S. Coast and Geodetic Survey projection experts Charles Deetz and Oscar Adams held the Mercator "responsible for many false impressions of the relative size of countries differing in latitude." These objections had no more impact than Erwin Raisz's 1938 explanation, in the first edition of his classic textbook *General Cartography,* that "the Mercator world map enjoys an unmerited popularity," perhaps because its areal exaggeration at higher latitudes helps mapmakers "represent the small countries of Europe on a world map." Must be able to locate Luxembourg and Liechtenstein, eh? However logical and strident, academic cartographers wrote for a small audience and had little influence on atlas publishers, textbook authors, or schoolteachers. According to Wong, the Mercator projection was so well established in the 1930s that most geographers considered other maps "unfamiliar and unconventional."

A stronger impetus for the Mercator map's decline was the war with Germany and Japan, which heightened public awareness of relative distance and cartographic perspective. Media critics mounted a pointed attack on the projection's scale distortions and exaggerated separation of the United States from Europe and Asia. Particularly influential was *Life* magazine's August 1942 illustrated essay "Maps: Global War Teaches Global Geography." Focusing on scale distortions and the relative advantages of conic and azimuthal representations, the nine-page exposé condemned the Mercator worldview as a "mental hazard in a war that is plotted on great circles across the land and sea and through the air." The following February a *New York Times* editorial declared that "the time has come to discard it for something that represents continents and directions less deceptively." Historian Susan Schulten, who chronicled the anti-Mercator campaign, discovered similar denunciations in periodicals as diverse as *Reader's Digest, Consumer Reports,* and the *American Scholar.*

Schulten attributed the Mercator's prewar preeminence to a "map industry [that] consciously chose to meet consumer expectations about the look and shape of the world." Hammond, Rand McNally, and their competitors believed they knew what the public wanted, while consumers conditioned by decades of "overexposure" to the Mercator framework trusted the publishers' expert judgment. The bland continued to lead the blind—or was it the reverse?—until the widening war encouraged radically different perspectives like the dramatic global snapshots drawn for *Fortune* by Richard Edes Harrison (1901–94). Trained as an architect, Harrison was a master of pictorial illustration and self-promotion. His 1944 atlas *Look at the World,* printed on cheap wartime paper, challenged conventional misrepresentations of the separation of Europe and North America with dramatic earth-from-space perspectives (fig. 9.2). In highlighting the disadvantages of mapping the globe onto a standardized north-up cylinder, Harrison cleared the way for fresh cartographic viewpoints. Enlightened by the media, clued-in consumers forced atlas makers to revise and retool.

To assess the endurance of Harrison's legacy, I checked out the world reference atlases on sale at my local Borders and Barnes & Noble. Although both stores offer a broad selection of atlases, hardbound and paperback, many of the books are reconfigured or abridged versions of an earlier, larger edition. In reporting the principal projection used for whole-world maps—a few atlases employ more than one—I list only the most recent version (table 9.1). As my table implies, none of the atlases exploited the Mercator framework for a world political or distribution map. What's more, the only instance of a Mercator whole-world reference map was the time-zone map in the otherwise progressive *Rand McNally Atlas of the World.* I can say this with confidence, even though nearly half the atlases failed to identify their projections by

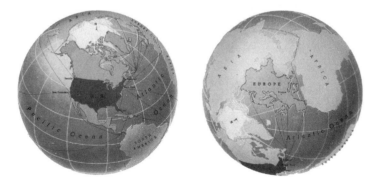

Figure 9.2 A two-page spread (52–53) in Richard Edes Harrison's *Look at the World* juxtaposed eight global views, two of which (shown here) boldly assert that the isolation of the United States (left) "is more seeming than real" and that Europe (right) has "more close neighbors than any other continent."

Table 9.1 Projections for whole-world maps in selected atlases

Atlas	World Projection
Barnes & Noble Essential Atlas of the World (2001)	[? (rectangular)]
Dorling-Kindersley Millennium World Atlas (1999)	Wagner VII
Hammond Citation World Atlas (2000)	Breisemeister
Hammond Compact Peters World Atlas (2002)	Gall-Peters
Hammond Concise World Atlas (2000)	Robinson
National Geographic Family Reference Atlas (2002)	Winkel tripel
Oxford Essential World Atlas (2001)	Hammer equal-area
Planet Earth Macmillan World Atlas (1997)	[Eckert IV?]
Rand McNally Atlas of the World (2001)	Robinson
Reader's Digest/Bartholomew Illustrated Atlas of the World (2001)	Eckert IV
Reader's Digest Illustrated Great World Atlas (1997)	Robinson
The Times Atlas of the World: Family Edition (1998)	Eckert IV

name, because the Mercator graticule is readily recognizable. Not so for the two projections listed in brackets, which I can only guess or describe vaguely. (Arthur Hinks also complained of unidentified projections in 1912.) Clearly, atlas publishers no longer see a need to cast world maps on a Mercator projection. Equally apparent is the diversity of world-map projections from which they can choose.

As my tabulation suggests, an abundance of more suitable world maps contributed to the Mercator's ouster. Of the projections listed, all but the Robinson and the Winkel tripel are equal area. But these two exceptions are compromise projections, designed to balance distortions of shape and area; neither misrepresents area as flagrantly as the Mercator. What's more, an atlas employing a compromise framework for large two-page reference maps often uses an equal-area projection for smaller whole-world distribution maps. Equally revealing, none of the replacements was available when the Mercator map became preeminent in the early nineteenth century.

A prominent compromise from the World War II era is the Miller cylindrical projection, designed by Osborn Maitland Miller (1897–1979), the American Geographical Society's projection expert, at the request of Samuel Wittmore Boggs, chief cartographer at the State Department. Concerned about public misinterpretation of shape distortion on equal-area world maps as well as area distortion on the Mercator projection, Boggs asked Miller to compare widely used cylindrical projections and recommend improvements. Constrained by Boggs's preference for a cylindrical framework, Miller reduced the Mercator's areal exaggeration by modifying map distance from the equator to the parallels. I call this the 80 percent solution because, for each parallel, he applied the standard Mercator formula to a latitude only 80 percent as large, and then divided the result by 0.8. A simple example clarifies the process. For the North Pole (90° N), located an infinite distance away from the equator on a Mercator map—and thus never shown—the adjusted calculation consists of finding the projected position for 72° N (80 percent of 90°) on the Mercator grid and then dividing by 0.8. As Tissot's indicatrix illustrates, the Miller cylindrical projection (fig. 9.3), although neither conformal nor equal area,

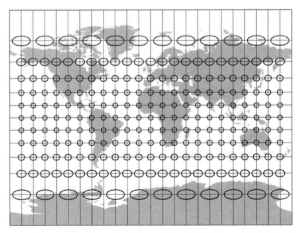

Figure 9.3 Modest angular distortion below 60° lets the Miller cylindrical projection show the poles and provide a Mercator-like treatment of equatorial and temperate regions. Miller's map found little use in Britain, where atlas publishers preferred an older, somewhat similar homegrown compromise projection, the rectangular Gall stereographic.

significantly lowers areal distortion for high latitudes. Although encompassing the whole world, including the poles, it mimics the Mercator map's familiar rendering of temperate and tropical regions.

Endorsed by the State Department and the American military, the Miller cylindrical projection made a modest contribution to the Mercator's decline. According to Frank Wong, the most significant inroads occurred in 1949, when the Miller map replaced the Mercator framework on the world political map in Rand McNally's *Cosmopolitan World Atlas,* and in 1951, when it displaced the familiar Mercator worldview in the *Encyclopaedia Britannica World Atlas.* A more lasting conquest was Rand McNally's adoption of the Miller grid for several climate maps in *Goode's School Atlas.* Substituted for the Mercator in 1949, it's still there in the most recent edition (2000).

Goode's World Atlas (as it is now called) is famous for another cartographic perspective: Goode's homolosine equal-area projection, developed in 1923 by University of Chicago geography professor J. Paul Goode (1862–1932), who moonlighted as Rand McNally's key cartographic consultant (fig. 9.4). Eager to preserve relative area but leery of the severe distortion of angles and shape on equal-area world maps, Goode divided the globe into six lobes, two above the equator and four below, interrupted over water to focus on land (fig. 9.5). Each lobe has its own central meridian, carefully positioned to minimize angular distortion over landmasses. To more faithfully imitate shapes portrayed on a globe, he subdivided each lobe into polar and equatorial zones, depicted respectively by Mollweide (homolographic) and sinusoidal projections that share the lobe's central meridian, around which distortion is minimal. Only in northeast Asia, well removed from both the equator and its lobe's central meridian, is a continent affected by pronounced shearing. By carefully tailoring his twelve zones and their regionally centered projections, Goode devised a remarkably realistic foundation for mapping population, vegetation, and other land-based distributions. And to further illustrate the benefits of interruption, he presented a second composite worldview with its continents split to reduce distortion over the oceans.

Cartographic textbooks treat the Mollweide and sinusoidal projec-

Figure 9.4 J. Paul Goode. Courtesy of
Robert B. McMaster.

tions (fig. 9.6) as "pseudocylindrical" modifications of the plane chart's
spartan rectangular framework. All pseudocylindrical projections
preserve relative area by bending meridians inward toward a central
meridian. Some treat the poles as lines shorter than the equator, while
others, like the Mollweide and the sinusoidal, map the pole as points.

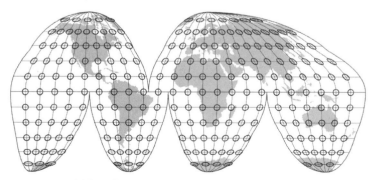

Figure 9.5 Goode's homolosine equal-area projection is a composite of twelve regionally centered projections. Knick points at 40° 44′ N and S mark the boundary between
polar and equatorial zones. Its overall shape suggests the skin of an orange, peeled
back and flattened.

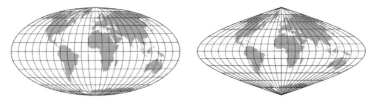

Figure 9.6 The Mollweide (left) and sinusoidal (right) projections minimize angular distortion in polar and equatorial zones, respectively, on Goode's homolosine equal-area projection.

Further adjustment in the spacing of parallels on the Mollweide reduces angular distortion around the poles and provided Goode with a low-shear solution for polar areas. By contrast, evenly spaced parallels on the sinusoidal afford a truer depiction of distance near the equator. The two projections form a seamless lobe by blending perfectly at 40° 44′ 11.98″ N and S, the latitudes at which their east–west scales are equal.

Goode's new world map did not grace his school atlas until 1932, when it replaced an interrupted Mollweide projection used since the atlas's debut in 1923. Educators eagerly accepted these segmented worldviews, perhaps because Goode not only integrated several more familiar projections, including the Mercator, but explained his strategy in the preface to the first edition: "It is quite impossible to transform the surface of a globe into a plane surface without the sacrifice of some elements of truth. It is not possible to have truth of angle, shape, area, and scale all in one map. . . . For geographic use, truth of area is of prime importance, and close to this is truth of form." To justify interrupted distribution maps, he cited four advantages:

(1) It presents the entire earth's surface, which Mercator's projection cannot do. (2) It is an equal area projection; there is no distortion of area. (3) Parallels of latitude are represented by straight lines trending with the equator, just as in the Mercator, a fine advantage in the study of comparative latitudes. (4) By the method of interruption of the grill [graticule], each continent in turn is given the advantage

of having a mid-meridian of its own; in this way better shapes are
given the continents than is possible with other projections.

Goode justified each of the atlas's other five projections, including
one incorporated largely to prove a point: "Only one Mercator projection is used in this atlas, and this is introduced so that its qualities may
be compared with the interrupted homolographic, the newer and better projection for a world map." After acknowledging that the Mercator's value to mariners "is so great that there seems to be no prospect
that any other projection will ever take its place for purposes of navigation," Goode vigorously attacked its misuse:

> In all previous atlases Mercator's projection has been used al-
> most exclusively for world distributions, and this in spite of the fact
> that (1) it is impossible with it to show the earth's surface entire, the
> north and south poles being at infinity; and (2) distances and areas
> grow rapidly larger with increase in latitude, becoming enormous in
> the higher latitudes. On a Mercator's map North America is much
> larger than Africa, although in fact North America is only seven-
> tenths the size of Africa. On a Mercator's map Greenland is larger
> than South America, when in reality it is only one-ninth as large as
> South America. This distortion of area is so bad that it becomes ped-
> agogically a crime to use Mercator's maps for studies of the relative
> sizes of continents and oceans, or for areal distribution of any kind.
> Population density, density of existing forests, annual rainfall, com-
> parison in size of states and empires, all are untrue and inexcusable
> as shown upon a Mercator chart.

I doubt that Rand McNally would have published a revolutionary
school atlas so unlike the firm's other products had Goode not earned
management's confidence through two decades of astute advice on
cartographic issues. Many employees disliked the diversity of projections and doubted the general public would appreciate an interrupted
world map. Resistance softened because of the school atlas's success
as a textbook and the cartographic turmoil of World War II, and the

homolosine projection quietly infiltrated the company's mass-market reference atlases. Even so, Rand McNally executives were still wary of interrupted world maps in 1961, when they asked America's leading academic cartographer, Arthur Robinson, to design a new projection. A key requirement was that the map not be interrupted.

Robinson's answer was a compromise projection that, like Miller's, blended distortions of angles and area but without the severe shear imposed by straight-line meridians. Keen to avoid "inducing lasting erroneous impressions, such as might result, for example, from the marked variation in area [on] Mercator's projection," he first canvassed existing projections for an acceptable solution. Finding none, he constructed a list of specifications that included relegating the inevitable exaggeration of area to polar areas, "even though this, unhappily, would greatly enlarge Antarctica." Relying on his sense of aesthetics—as an undergraduate he had studied both art and geography—Robinson employed the computer as a design tool in a trial-and-error search for a world map with a realistic look. After repeated rounds of plotting, appraising, and adjusting the world's shorelines, typically by tweaking the spacing and relative length of the map's parallels, he settled on a design initially named the orthophanic (right-appearing) projection. As Tissot's indicatrix shows in figure 9.7, the trade-off of areal and angular distortion, chiefly apparent in upper latitudes, is less pronounced than on projections that present the poles as points or stretch them to lines as long as the equator. Like other pseudocylindrical transformations, the Robinson projection favors landmasses close to its central meridian but affords a comparatively convincing treatment of Australia, central Asia, and most of North America. Its prominently rounded ends remind viewers that the only true world map is a globe.

Rand McNally unveiled the new map in 1965 as a wall map and in several mass-market atlases. Public response was discouraging—consumers apparently preferred the Mercator map—and Robinson's compromise was relegated to school atlases and similar educational products. At least that's the story circulating among academic cartographers. I've seen no evidence that the company assertively pitched its new projection to a general audience, and whatever market research it

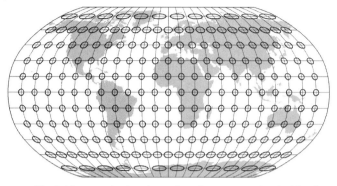

Figure 9.7 The Robinson projection, the product of computer-assisted trial-and-error refinement, reflects Arthur Robinson's effort to create a "realistic," uninterrupted map of the world.

carried out was never fully disclosed. But I also discovered the four-decades old projection alive and well in recent Rand McNally world atlases as well as in similar products from several competitors. What gives?

What happened reminds me of a pair of three-year-olds in a playgroup: one discards a toy, the other picks it up, and the first wants it back. In this parable the first kid is Rand, the second has the initials NGS, and the toy is Robinson's map, which the National Geographic Society gleefully picked up in 1988, touted in a flurry of press releases, and discarded a decade later for the Winkel tripel (or Winkel III) projection, a modified azimuthal map introduced in Germany in 1921 by Oswald Winkel (1873–1953) but rarely used. Like the Robinson map and the Van der Grinten projection (fig. 9.8, left), which framed the official NGS world map between 1922 and 1988, the Winkel tripel is a compromise perspective, neither equal area nor conformal (fig. 9.8, right). Its quiet introduction in fall 1998 was subliminal in comparison to the Robinson's hyped debut ten years earlier: although the Society mailed a free, 4 by 6 foot laminated world map to every public and private school in the United States and Canada, few news stories mentioned the map's Winkel tripel framework, named but not described in the eighth paragraph of a thirteen-paragraph press release.

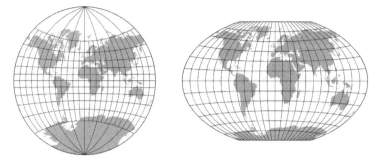

Figure 9.8 The Van der Grinten projection (left) framed the National Geographic Society's official world map from 1922 to 1988, when it was supplanted by the Robinson projection, itself replaced in 1998 by the Winkel tripel projection (right).

Am I suggesting that major map publishers have the attention span of a three-year-old? Perhaps, but the Society's 1988 adoption of the Robinson map was hardly impulsive. I know because I participated in the December 16–17, 1987, seminar convened at NGS headquarters in Washington, D.C., to help chief cartographer John Garver select the best projection for its world reference map. In addition to Garver and six other NGS officials, the discussion included John Snyder, me, and two other academic cartographers, Judy Olson and Dick Dahlberg. After dissecting numerous contenders, we unanimously endorsed the Robinson projection. A year later—designing and printing millions of map supplements doesn't happen overnight—*National Geographic* magazine hit the streets with a large folded freebie. Garver's accompanying article, "New Perspective on the World," praised the new map's "different and more realistic view of the world," judged it "better still" than "the trusty Van der Grinten," and "hope[d] that its main legacy will be a generation of map readers more critical of all flat maps." Inspired by an NGS press release, more than 550 newspapers and magazines with a total circulation greater than 51 million reported the changeover. Eat your heart out, Rand McNally.

National Geographic publications used other global perspectives but flaunted the Robinson projection as the Society's signature world map. In 1990, for example, the newly revised sixth edition of the *Na-*

tional Geographic Atlas of the World featured a large foldout with back-to-back world political and world physical maps, both cast on Robinson's projection. The lower right corner of the political map reported the projection's adoption in 1988 and claimed that it "presents a more realistic view of the world." This begged the question, "more realistic" than what?

Realism lost its luster in less than a decade. In 1999 the atlas's seventh edition examined map distortion in an impressive two-page treatment titled "Round Earth, Flat Paper." Conveniently ignoring the Mercator projection's role in framing *National Geographic* magazine's first world map supplement, inserted in the February 1905 issue, an illustrated sidebar on the "Evolution of a Better World" chronicled the Society's successive use of azimuthal equidistant, Van der Grinten, Robinson, and Winkel tripel perspectives. The Robinson map, we're told, shows "more than 75 percent of the Earth's surface . . . with less than a 20 percent departure from its true size and scale," while its replacement "avoids the congestion and compression of higher latitude areas and lessens distortion of scale and shape." Readers baffled by this verbal geometry—no, you're not alone—can rely on the not-so-subtle Darwinian analogy, which leaves little doubt about the Winkel tripel's representational superiority.

If nothing else, a new cartographic signature is an occasion for proclaiming a publisher's commitment to progress, much like the laundry detergent proud to be "new and improved" or the retailer obsessed with reconfiguring aisles, shelves, and signage. I call this the Monty Python Effect, after the comic troupe's famed transition line, "And now for something completely different." National Geographic does it, Rand McNally does it, and eventually they all do it, at least partly because the trade-offs inherent in map projection make doing it easy and effective.

A new mathematical formula is not the only way to radically alter a world map. Where the mapmaker centers a projection can enhance a map's distinctive look as well as privilege some places by making them less peripheral, and thus more obvious if not more important than places near the edge or, worse yet, split in two. Until 1975, when it recentered its world map on the Greenwich meridian (0°), the National Geographic Society copied nineteenth-century United States atlas publishers who favored the ninetieth meridian, a fraction of a degree west of Chicago and especially convenient for its rival Rand McNally, headquartered in the Windy City. Although the America-centric tradition is alive and well in Mercatorized world wall maps advertised on the Internet (as in fig. 9.1), most United States atlas makers now center their world maps on Greenwich, probably because splitting Tibet and severing Siberia from the rest of Russia are aesthetically awkward if not geopolitically inappropriate. Far better to interrupt the cylinder at the 180th meridian and show North America and Asia as whole continents by repeating small amounts of Alaska and Siberia in the empty upper corners of a pseudocylindrical projection. Clever redundancy can improve the look and help the user.

Some projections invite cropping. For example, National Geographic cartographers found it necessary to sever upper and lower portions of the Van der Grinten projection (fig. 9.8, left), which mapped the entire world into a circle but squandered space on polar areas with few features worth naming. Cropping was seldom perfectly symmetric: when a label near the top required additional room, they'd lop a bit more off the bottom.

Mercator's projection makes cropping mandatory. As figure 9.9 illustrates, a Mercator map truncated at 89° N and S portrays Antarctica as larger than all other continents combined. Some of the southern continent must go, of course, but how much? Barring a need to show specific features in polar zones, cartographers are influenced largely by the ratio of width to height, which determines whether a map looks good and fits the page. In opting for an aesthetically acceptable format, they pass up a dramatic demonstration of the Mercator's outrageous areal exaggeration. Less artistic motives might also be at work: cutting the bottom of the map off at the tip of South America while showing most of Greenland puts western Europe closer to center stage—a worldview now widely condemned as "Eurocentric." But un-

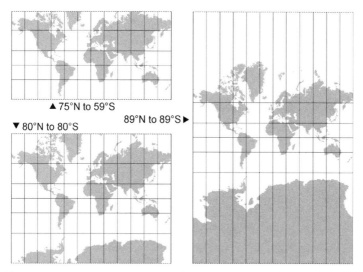

Figure 9.9 How a Mercator map is cropped can greatly affect its worldview and aspect (width/height) ratio.

less coverage extends well north of Greenland, the map's center typically winds up in north or central Africa.

Is a Europe-centered map subliminally malevolent or merely a natural accommodation of European readers? In some situations centering a map near the viewer's location is good design. But if you're a humanist well versed in European imperialism's harsh imprint on the Third World, the traditional Greenwich-anchored world map becomes a clear example of Western cultural hegemony, and all the more so when a Mercator projection inflates the size of western Europe. Outraged humanists must, of course, overlook the greater (but imperially irrelevant) prominence of Siberia, northern Canada, Greenland, and perhaps a lot of Antarctica. More clearly problematic is the historical atlas that largely ignores Asian or Islamic civilizations or the world atlas with ostensibly racist overtones in its themes, regional groupings, and categories. Eurocentric cartography can be far more shameful than a Greenwich-anchored world map.

An easy target is the early-twentieth-century world map that re-peats Australia in its lower left and lower right corners. London jour-nalist Simon Jenkins recently recalled such a map from his boyhood, in a 1926 textbook "long used in British schools after the war." Cen-tered on 40° W, its Mercator projection "made Canada look far bigger than the United States and depicted in red the huge British claims in Antarctica." This arrangement clearly exaggerates the size and extent of the British Empire, but its role in promoting a sense of superiority is debatable. (If the colonial Brits' self-esteem required a cartographic recharge, they're hardly as arrogant as we're led to believe.) Although two Australias reinforced the now-quaint notion that the sun never sets on England's possessions, this redundancy also affords uninter-rupted treatment of Oceania to the left and South Asia to the right. Both interpretations (cultural hegemony and effective design) are valid, and I'm not certain it's worth debating which one is more fun-damental.

It's surprisingly easy to read unintended meaning into an other-wise innocuous world map. Consider the George F. Cram Co.'s *Map of the World* that I picked up recently at a cartographic meeting. An-chored on 90° W, the asymmetrically cropped map is centered ver-tically in northern Mississippi or western Tennessee, perhaps on Memphis, which would make it an Elvis-centered world map and thus a clever symbol of American cultural imperialism. OK, Elvis centering is a straw man. I'm not denying American hegemony—cultural or otherwise—or its sometimes sinister implications, but I doubt this map has anything to do with Americans' views of where we can ped-dle schlock or bomb with impunity. Centering a map midway between Greenwich and the International Date Line might be more no more meaningful than a cultural preference for rounded numbers like 50 and 100.

If cartographic scholars want to puff up the importance of the ar-tifacts we study, and raise our profile in the process, cultural hege-mony clearly makes an attractive explanation. How tempting to claim, as the distinguished cartographic historian Brian Harley once did, that "the simple fact that Europe is at the center of the world . . . must have contributed much to a European sense of superiority." But can this argument and its intellectual kinfolk survive the "So what?" test? It's a question we seldom ask.

Similar "maps matter" explanations are equally suspect. A Chicago-anchored Mercator map could have reinforced Americans' sense of isolation from Europe in the late 1930s and early 1940s, but did a map in any measurable way delay our country's entry into World War II? And even though a map with Japan and Hawaii on opposite sides of the world argued against a Japanese attack on Pearl Harbor, is there any evidence, archival or otherwise, that an America-centered projection either stifled military intelligence in early December 1941 or deterred strategic planning in the months before? If so, it's a well-kept secret.

Blind faith in the power of maps comes easy when a projection suggests or magnifies a threat. Because a naive public accepts maps as facts, the John Birch Society speaker delivering an anti-Communist diatribe in the 1960s and 1970s often shared the stage with a large Mercator display, with the Soviet Union, China, and assorted client states colored a brilliant, threatening red. Proponents of air power and missile defense found the markedly different viewpoint of an azimuthal projection useful in dramatizing the possibility of an over-the-pole attack from the Soviet Union. And after the Communist Bloc disintegrated in the late 1980s, the Robinson projection, as Ben Wattenberg observed, "out-perestroika[ed] perestroika." Map historians who gleefully celebrate these alleged cartographic contributions to the Cold War might usefully ask whether the map's role is a matter of intrinsic power or merely the convenient availability of diverse designs.

Some maps might matter. A classic case is Sir Halford Mackinder's (1861–1947) theory of the Heartland, unveiled in 1904 at a meeting of the Royal Geographical Society. An academic geographer with political ambitions, Mackinder believed that technology would soon make ships less important than railways in controlling energy and food resources. A "pivot area" extending from eastern Europe into Russia and south into the Middle East enjoyed a natural advantage, he argued, and "its expansion over the marginal lands of Euro-Asia would permit

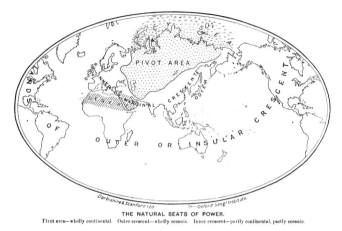

Figure 9.10 Sir Halford Mackinder promoted his Heartland theory with this geopoliti-cally suggestive oval framing of a Mercator map. From Mackinder, "Geographical Pivot," 435.

the use of vast continental resources for fleet-building [and] the em-pire of the world would then be in sight"—an imminent threat "if Ger-many were to ally herself with Russia." To dramatize his argument, Mackinder devised a Mercator map (fig. 9.10) on which duplicate Americas and an oval frame imply an invincible "heart-land" sur-rounded by an inner crescent of partly continental states and an outer crescent of "wholly oceanic" states, disadvantaged by the emerging im-portance of land transportation. Mackinder's map is important be-cause Karl Haushofer, a German political geographer who advised Adolf Hitler, bought the idea that control of Eastern Europe was the key to world domination.

However influential Mackinder's elliptical map, its projection merely reflected public perception of how continents should look on a world map. Areal exaggeration was not an asset: Eurasia was a natural Heartland, and Mackinder needed an oval frame to downplay the artificial inflation of the theoretically peripheral Canada and Green-land. The equal-area Mollweide framework would have worked just as well, if not better, had it been the established ideal. What the Mercator worldview offered was nothing more (or less) than authenticity. Through the mid-1940s, it was the undisputed standard, which Miller and Van der Grinten felt compelled to partly replicate. In the absence of another powerful "master image," as Peter Vujakovic notes, the Mercator projection remains more widely used than its unique navigational qualities warrant.

Size Matters

High school debating coaches frown on misrepresenting an opponent's position as a straw man argument, easily knocked over like a dummy filled with straw. Instantly recognized by debaters and judges alike, this rhetorical tactic is common in advertising and political debate, where audiences are less likely to recognize logical flaws, and in the news media, where reporters eager for a lively controversy accord both sides equal time in the guise of fairness. It works especially well when your opponent's counterargument is too technical for general readers. Because any attempt to sort out the facts might be interpreted as taking sides, journalists typically treat both positions as equally valid, as they did when advocates of the Peters projection promoted their "revolutionary" world map as an antidote for the Mercator projection's dastardly "Eurocentric" worldview.

Battle lines were clearly drawn, with the Mercator map on one side and the Peters map on the other. In condemning the former as detrimental to Third World nations, which are largely in the tropics and thus downsized relative to Western Europe, the United States, and the

145

developed world in general, Peters's proponents ignored decades of pointed criticism of the Mercator map by academic cartographers and its wholesale abandonment by map publishers in the 1940s. Their simplistic scenario linked the Mercator map to an unsympathetic cartographic establishment and aligned the Peters map with pro–Third World organizations like Oxfam and UNESCO—credible allies, even if their expertise has no bearing on the issue. Not content with mere endorsement, the World Council of Churches and similar organizations implied that to oppose the Peters projection was to support intolerance and economic exploitation. To help their readers understand the controversy, newspapers and magazines juxtaposed examples of the Mercator and Peters world maps (fig. 10.1)—as if these were the

rect sizes. To do so, it enlarges and elongates most Third World countries at the expense of the northern hemisphere, particularly Europe. That's exactly what Peters, a historian from Bremen, West Germany, had in mind when he drew the map.

Another German, Gerhard Kramer, first drew the more familiar Mercator map in 1569. (Kramer Latinized his last name to "Mercator.") His map produced severe size distortions in some countries because he located his own homeland, Germany, in the center of the map. It

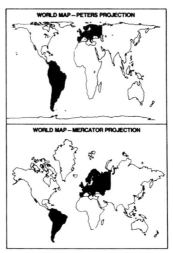

The Peters projection, top, reflects the actual sizes of land masses. The Mercator map, bottom, makes Europe appear larger than South America. It is actually smaller.

although it distorts shapes in order to preserve accurate geographical relationships.

"In our epoch, relatively young nations of the world have cast off the colonial dependencies and now fight for equal rights," he says. "It seems important to me that the developed nations are no longer placed at the center of the world, but are plotted according to their true size." His work is perhaps more of a contribution to world politics than it is to cartography, since its shape distortion renders it unsuitable for use in navigation. *(cont.)*

Figure 10.1 *Christianity Today* illustrated a pro-Peters story with this pointed comparison of the Peters and Mercator portrayals of Europe and South America. From "A New View of the World," 39.

only choices—and used visual propaganda to put professional cartographers on the defensive.

Who was Arno Peters? According to an obituary that appeared in the *Times* of London shortly after his death at age eighty-six on December 2, 2002, he was a German historian recognized as a staunch "advocate of equality in all things." In addition to his map projection, Peters and his first wife, Anneliese, authored the *Synchronoptische Weltgeschichte* (Synchronoptic World History), an elaborate book-length, year-by-year timeline that runs from 1000 BC to AD 1952 and devotes equal space to all years—a bafflingly obsessive strategy in light of the vast differences between earlier and later years in numbers of inhabitants, inventions, and noteworthy events. Other Peters innovations include a system of musical notation that distinguishes notes by color as well as position on the scale and a postcolonial global coordinate system based on decimal degrees and a prime meridian aligned with the International Date Line. Noting that the Greenwich meridian became the global standard in 1884, "when Britain was the strongest European colonial power and ruled over a quarter of the world," he argued that with "the ending of colonialism and the closure of Greenwich Observatory, there is no reason other than custom for retaining this zero meridian."

Peters came to cartography late in life. Born in Berlin in 1916, he studied journalism, history, art, and film production. After receiving a doctorate in 1945 from Berlin's Friedrich-Wilhelm University—his dissertation, "Film as a Means of Public Leadership," seems especially timely—he secured various grants, including one from the American government, for a world history textbook suitable for use in both East and West Germany, neither of which was particularly pleased with the result. After writing for a socialist magazine from 1958 to 1964, he co-founded the Institute for Universal History in Bremen and became its director in 1975. Although Peters discussed his world map as early as 1967, at a meeting of the Hungarian Academy of Science, he didn't actively promote it until May 1973, at a press conference in Bonn. Reporters received copies of his *Orthogonal Map of the World* and a brochure, *The Europe-Centered Nature of Our Geographical Picture of the World and Its Conquest.* As the brochure's title implies, Peters used his ostensibly egalitarian world map to condemn colonialism and Western dominance.

Academic cartographers who might have endorsed the map's political message challenged the projection's appearance and purported advantages. Among the earliest critics was Derek Maling, a map projection expert at the University of Swansea, in Wales. In the *Geographical Journal* Maling called the brochure "a remarkable example of sophism and cartographic deception" founded on "the time-honoured pursuit of denigrating the value of Mercator's projection as the base for a world political map." In *Geographical Magazine,* he quoted Karl Heinrich Wagner (1906–85), a German mathematical cartographer who a few months earlier had opined, "The whole ten-year wonderwork could be accomplished in ten minutes with the aid of a little elementary arithmetic." And in *Kartographische Nachrichten,* published by the German Cartographic Society, he chided Peters for "the remarkable discovery that the Equator on Mercator's Projection is not located in the middle of the map so that about 2/3rds of the map shows the Northern Hemisphere and only 1/3rd represents the Southern Hemisphere"—a revelation conveniently confirmed "by a map on Mercator's Projection in which the two limiting parallels happen to be located in latitudes 80° North and 60° South." To make the Mercator map Eurocentric, it's essential to cut off Antarctica while showing all of Greenland.

Sarcasm was only part of Maling's attack. In addition to dismissing the professed novelty of the Peters projection, he demolished the absurd claim that it accurately represented all distances—as mathematicians and cartographers are well aware, no flat map can do that. For his coup de grâce, Maling resorted to his specialty, cartometry (the science of making measurements on maps), to demonstrate that the Peters projection was not even, as claimed, an equal-area map. Close inspection revealed that the parallels on the version presented at the Bonn press conference were off by as much as four millimeters from their correct positions on a true equal-area map at the same scale. Although four millimeters might not seem a large discrepancy, it was, in terms British readers might best appreciate, "equivalent to shifting Bristol to the latitude of the Channel Islands or London to the latitude of Paris."

Believing that Peters had erred, Maling proposed two explanations: either the historian had made mistakes in drawing an equal-area cylindrical projection designed with standard parallels at 46° 02' N and S, or he had more seriously botched an equal-area projection intended to have standard parallels at 45°. A revised map that Peters released in 1975 confirmed the second explanation. According to Maling, Peters "confessed" in 1980 to drawing the 1973 map incorrectly and correcting his mistake in 1975.

Among academic cartographers this latter explanation raises questions of priority and independent invention—serious questions to academic scientists like Maling. Arno Peters was not the first to present an equal-area cylindrical projection secant at 45°. James Gall (1808–95), a Scottish clergyman, described an identical projection in 1855 at a meeting of the British Association for the Advancement of Science and discussed it more fully in 1885 in *Scottish Geographical Magazine.* His article, which introduced two other cylindrical projections, included a picture of an equal-area map he called Gall's Orthographic projection (fig. 10.2). Gall didn't think much of his equal-area map, on which "geographical features are more distorted . . . than on [the other two] but . . . not so distorted as to be unrecognizable." He recommended it largely "for showing the comparative area occupied by different subjects, such as land and water, as well as many other scientific and statistical facts."

A Victorian gentleman scientist, Gall altruistically donated his three projections to whoever wanted them. "All I would ask," he implored, "is that, when they are used, my name be associated with them, and that they be severally distinguished as Gall's Stereographic, Isographic, and Orthographic Projections." Of the three, only Gall's Stereographic gained wide use in British atlases, for which it provided an alternative to the Mercator projection by balancing distortions of area and angles. By contrast, his equal-area offering (Gall's Orthographic) was rarely used or mentioned, apparently because of shapes too dis-

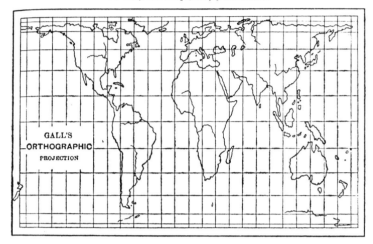

GALL'S ORTHOGRAPHIC PROJECTION.

EQUAL AREA. PERFECT.

For Physical Maps, chiefly Statistical.

GALL'S
ORTHOGRAPHIC
PROJECTION

Figure 10.2 Gall's Orthographic projection, a forerunner of the Peters projection and prototype for the Gall-Peters projection. From Gall, "Use of Cylindrical Projections," 121.

torted for nineteenth-century atlas publishers, who typically turned to a sinusoidal or Mollweide projection whenever they needed an equal-area map.

Could Peters have discovered his projection independently? Sure. Is it likely? Yes, but only if Peters naively ignored the existing literature on map projection, including a 1910 German-language article by Walter Behrmann, who pointed out that giving a cylindrical equal-area projection two standard parallels, rather than one (the equator), reduced the pronounced east–west stretching of polar areas on the original rectangular equal-area map, presented by J. H. Lambert in 1772. Curious about the overall effect on shape, Behrmann examined angular deformation for rectangular projections secant at latitudes between 10° and 60°, at a ten-degree interval, and concluded that standard parallels at 30° yielded "the best of all known equal-area projections of the whole world."

"All known," it turns out, is a very large group (fig. 10.3) insofar as every marginally distinct pair of standard parallels produces a slightly different world map. It helps, though, to assert broad benefits like those claimed by British architect Trystan Edwards for a version secant at 37° 24′, presented in 1953. Edwards promoted his projection "as a 'general service' map to supercede Mercator's for all purposes except navigation and as the standard world map for the study of political geography." A *Times* of London editorial titled "Mercator Disci-

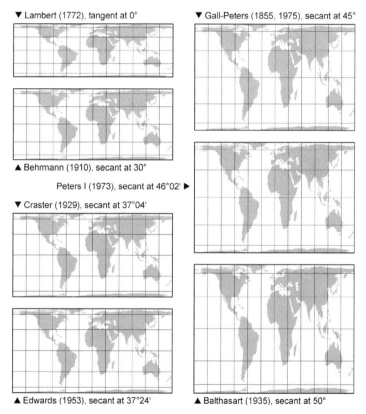

▼ Lambert (1772), tangent at 0°   ▼ Gall-Peters (1855, 1975), secant at 45°

▲ Behrmann (1910), secant at 30°

Peters I (1973), secant at 46°02′ ▶

▼ Craster (1929), secant at 37°04′

▲ Edwards (1953), secant at 37°24′   ▲ Balthasart (1935), secant at 50°

Figure 10.3 A few members of the family of cylindrical equal-area projections, a very large group insofar as every plausibly distinct pair of standard parallels defines a new member.

plined" puts him a couple of decades ahead of Peters in media-savvy Mercator bashing. Optimistic but myopic, Edwards patented his projection, thereby assuring its obscurity—why pay a royalty, mapmakers ask, when equally sensible options are free for the taking? Other named members of the cylindrical equal-area family are the Craster rectangular equal-area projection (secant at 37° 04′), mentioned fleetingly in a 1929 article by Colonel J. E. E. Craster, who much preferred its pseudocylindrical modification, and the equally obscure Balthasart projection (secant at 50°), introduced in 1935 in a Belgian geographic journal by the otherwise unknown M. Balthasart. Jeremy Crampton, an academic cartographer eager to give Peters the benefit of the doubt, would also include the German historian's admittedly flawed 1973 map (secant at 46° 02′).

Although Peters was probably unaware of the Gall Orthographic projection in the mid-1970s, he should have done a bit more homework before touting its advantages in an impressively designed, generously illustrated book, *The New Cartography,* published in 1983. Not content with "fidelity of area," he made claims that were either false (like "fidelity of scale," possible only on a globe) or trivial (like "totality," which describes a map's ability to show the entire world). Sure, the Mercator and a handful of other projections can't cover the entire sphere, but no equal-area projection lacks totality. A few of his "attainable map qualities" seem contrived to exclude interrupted composite equal-area projections like Goode's (see fig. 9.5), which lacks "fidelity of axis" because its meridians curve, "supplementability" because the mapmaker can't "detach a [small] section from the left hand side and . . . reattach it to the right," and "proportionality," which demands "longitudinal distortion along [the map's] upper edge as great (or as small) as along its lower edge." Eager to disparage a superior alternative, Peters unfairly denied Goode's map "fidelity of position," whereby "all points which exist at an equal distance from the equator are portrayed as lying on a line parallel to the equator." Goode's parallels might be interrupted, but they're consistently parallel to the equator.

Three additional qualities round out Peters's list of ten. Willingly conceding that Goode's, Mercator's, and Van der Grinten's projections (among others) possess "clarity," whereby a map "does not deform by extreme distortion any of the countries, continents and seas portrayed," Peters boldly claimed this attribute for his own map. And technically he's right: its strung-out continents are not "extreme" insofar as their distortions could, technically, always be worse. The two remaining qualities are vague and baffling. The Peters projection has "universality" because it "permits the construction of grid systems for maps of each section of the earth's surface as well as for a global map ... and permits the portrayal of all contents of a map for all applications." Huh? And because of "adaptability" it "can cope with specialist requirements of general map contents." According to these enigmatic definitions, no projection other than Peters's is universal or adaptable.

To the chagrin of the German Cartographic Society, which challenged Peters's claims as early as 1973, the assertive historian won the support of former German chancellor and Nobel laureate Willy Brandt, who chaired the Independent Commission on International Development Issues. To advertise disparities between developed and less developed nations, the latter largely in the tropics, the Brandt Commission put Peters's map on the front covers of its 1980 report *North–South: A Programme for Survival* and a 1983 sequel, *Common Crisis North–South: Cooperation for World Recovery.* Identical acknowledgments note the projection's use "rather than the more familiar Mercator projection," mention its "several innovative characteristics," and praise it for "represent[ing] an important step away from the prevailing Eurocentric geographical and cultural concept of the world." Prominently endorsed, the map that promised equality and fairness became an icon of social consciousness.

Relief organizations and other pro–Third World groups worldwide began using the Peters map in their own publications or giving huge numbers of wall-size copies to schools and churches. According to the *Economist,* by 1989 UNICEF and kindred agencies had distributed over 60 million copies. Prominent adopters include the United Nations Educational, Scientific and Cultural Organization (UNESCO) on the international scene and the National Council of Churches of Christ in the United States. In Britain there's Action Aid, Oxfam, and the Third World Foundation.

Why the Peters map? In 1986 geographer Peter Vujakovic surveyed British groups committed to Third World development and found that twenty-five of forty-two respondents (69 percent) had adopted the Peters projection for one or more publications. Asked to explain their decision, adopters cited a variety of reasons, including areal equality (48 percent), a distinctive look that commands attention and provokes thought (36 percent), elimination of Eurocentric bias (32 percent), and improved recognition of Third World countries (24 percent). Despite its distortion, the map's unusual appearance was clearly an asset. As one respondent noted, "use of the Peters projection is a statement in itself."

In framing his map as the only effective antidote to a venomous Eurocentric cartography, Peters appealed to adopters' wariness of Western values. The enemy was not just Eurocentric maps, he argued, but the professional mapmakers who foisted them on an unwitting public. Calling for a revolution in mapmaking, he attacked cartographers' authority, ethics, and relevance.

> Philosophers, astronomers, historians, popes and mathematicians have all drawn global maps long before cartographers as such existed. Cartographers appeared in the "Age of Discovery," which developed into the Age of European Conquest and Exploitation and took over the task of making maps.
>
> By the authority of their profession they have hindered its development. Since Mercator produced his global map over four hundred years ago for the age of European world domination, cartographers have clung to it despite its having been long outdated by events. They have sought to render it topical by cosmetic corrections.
>
> . . . The Eurocentric world concept, as the last expression of a subjective global view of primitive peoples, must give way to an objective global concept.
>
> The cartographic profession is, by its retention of old precepts based on the Eurocentric global concept, incapable of developing this egalitarian world map which alone can demonstrate the parity of all the peoples of the earth.

Academic cartographers who might have been mildly amused by Peters's 1973 posturing were outraged and embarrassed by his assault on their credibility. Particularly offensive was the 1977 republication, apparently verbatim, of a West German government press release in the *Bulletin* of the American Congress on Surveying and Mapping (ACSM), a broad professional society in which land surveyors greatly outnumber cartographers. The *Bulletin*'s editor, a communications professional with no cartographic training, apparently welcomed the short piece, engaging and well illustrated, as a bone to throw members clamoring for cartographic content.

It's easy to see why a reader unfamiliar with map projection would find the item intriguing and convincing. Its two-part headline sets up the straw man with a plea for fairness:

FOUR CENTURIES AFTER MERCATOR:
Peters Projection—to Each Country Its Due on the World Map

A catchy lead sentence alludes to a pervasive problem:

The picture of the world we still use today originated in the world of yesterday.

Ignoring the great mapmaker's original depiction, centered slightly off the west coast of Africa (see fig. 4.2), a single short paragraph claims Mercator's genius for Germany while declaring his wall map Eurocentric:

Mercator had put his chosen homeland, Germany, lying in the northern quarter of the globe, in the center of the map together with all of Europe.

Areal distortion, the reader learns, makes the map racist:

The regions of the globe inhabited by white people were depicted much larger than the others. Countries and continents inhabited by colored people appeared a great deal smaller by comparison than they actually were.

Fortunately, another German discovered a solution a few years ago:

> Dr. Peters presented his new map to the world in 1973. It shows all countries and continents in their correct relative proportions. This absolutely area-factual Peters Projection furthermore keeps the unavoidable distortions of forms, distances, and angles so minimal that a world picture of great faithfulness to reality came into being.

Whatever credibility the piece gained among savvy readers by acknowledging the "unavoidable distortions of . . . distances and angles" was lost a few paragraphs later with unattainable claims that Peters's equal-area map "achieves absolute angle conformality" and "is totally distance-factual." Equal-area maps are never conformal.

American cartographers fired back in the next issue. Arthur Robinson, a professor at the University of Wisconsin and the author of a popular cartography textbook as well as his own world map projection (see fig. 9.7), questioned Peters's originality and common sense: "Map projections are fascinating for many reasons, not the least of which is the way people, such as Dr. Arno Peters, who know little of the subject, regularly devise something new and wonderful. Some of these 'discoveries' pass into oblivion because the originators have the good sense to check out the idea with the cartographer knowledgeable about projections. Others don't have such good sense, don't realize what they don't know, put it forward, and end up looking ridiculous. 'Peters' Projection' is a good example. Let me analyze it." Robinson, who would become the historian's fiercest critic, regarded Peters's map as nothing more than a useless modification of the cylindrical equal-area projection with two standard parallels. Behrmann's version, secant at 30°, had some merit, but not Peters's. "Most forms," he noted, "will appear either to squash the land areas N-S or stretch them." Robinson called some of Peters's claims "ridiculous" and argued that "only one who is blind could say that [his map] 'has no extreme distortion of form.'"

Following Robinson's commentary, projection expert John Snyder contended that the *Bulletin* article "reads much more like an exaggerated advertisement than like a professional presentation." Snyder
noted that Lambert had presented a cylindrical equal-area map in
1772 and that the Edwards and Behrmann variants "were also advanced as revolutionary." Citing Maling's analysis in *Geographical Magazine,* he questioned whether the Peters projection was, as claimed,
an equal-area map. His parting shot reflected frustration over what
seemed an obstinate ignorance of map history: "For Peters' promoters
to declare that this is the first world map projection of consequence
since Mercator is ridiculous and insulting to dozens of other inventors
over the years who have done a much better job with much more innovation and much less fanfare." The *Bulletin*'s editor washed her
hands of the flap with an optimistic note of closure: "The Dr. Arno Peters' projection was acclaimed by an F.R.G. Government release as having 'radically changed' the world map. Our eminent cartographers
have debunked the projection and seem to have 'laid it to rest' forever
in no uncertain terms!"

Laid to rest? Hardly. A decade later ACSM and Peters's supporters
were at loggerheads over a forty-two-page Friendship Press booklet
that misconstrued the *Bulletin*'s republication of the West German
press release as an endorsement of Peters's map. Friendship Press is
the publishing arm of the National Council of Churches of Christ in
the USA, and its executive director at that time was Ward Kaiser, an or-
dained minister in the United Church of Canada. In 1987 the Press
published Kaiser's *A New View of the World,* subtitled *A Handbook to
the World Map, the Peters Projection.* The contested passage, below, followed a brief listing of supporting organizations.

> Support for Professor Peters' map has been forthcoming from a
> number of professional communities. Geographers and cartographers
> among these. Thus the American Congress on Surveying and Map-
> ping could say: "[Dr. Peters' map] shows all countries and continents
> in their correct relative proportions. This absolutely area-factual
> Peters Projection furthermore keeps the unavoidable distortions

of forms, distances, and angles so minimal that a world picture of great faithfulness to reality came into being."

An endnote attributed the quote to the November 1977 issue of the *Bulletin* but omitted its origin as part of the FRG press release. Twenty pages later Kaiser concluded a discussion of the map's alleged qualities by emphasizing, "Thus the mathematical or scientific superiority of this projection becomes apparent. You may wish to quote statements that recognize this breakthrough, such as those by the American Congress on Surveying and Mapping."

ACSM objected vehemently and demanded a retraction. Clearly embarrassed, Friendship Press responded by pasting a short note on the contents page of unsold copies:

CORRECTION

The statement attributed to the American Congress on Surveying and Mapping on page 10, also referred to on page 31, was originally made in a bulletin of the Press and Information Office of the Government of the Federal Republic of Germany. The ACSM reprinted the material in its Bulletin No. 50 in November 1977 but has not made an official statement of its own on the Peters Projection.

As Peters's chief North American apologist, Kaiser had sparred with Robinson, Snyder, and other leading cartographers in countless articles and interviews. "Every teacher who really wants to help his or her students understand the real world ought to have this map," Kaiser told *USA Today*. Reiterating the Peters straw man for *Mother Jones*, he asserted that the Mercator map "makes the predominantly white-dominated areas of the world seem more important than they are." The same article includes John Snyder's observation that "no flat map will do perfect justice to a globe." Like other cartographers who rejected Peters's claim of priority, Snyder referred to the historian's creation as the "Gall-Peters projection."

Snyder considered both shape and area important. In an article for *Christian Century*, he recalled a radio interview in which the host had pressed Kaiser on Peters's distorted rendering of Africa. The minister's reply suggested ignorance of globes and space imagery: "Well, one needs to ask what is the normal shape of Africa? Without having seen Africa from outer space, I'm not really in a very good position, nor perhaps [is] any of us, to say how it actually looks." Phooey, Snyder retorted, "We know very well how Africa looks from space!" And "Peters' distortion of Africa seem[s] excessive." How excessive? In a "views and opinions" essay for *The American Cartographer,* Arthur Robinson likened the map's distorted landmasses to "wet, ragged, long, winter underwear hung out to dry on the Arctic Circle."

Robinson and Snyder felt the profession should take a stand. In 1985 the American Cartographic Association, the cartographers' wing of ACSM, formed a Committee on Map Projections, which examined the issue of distortion on world maps and produced three booklets explaining options to a lay audience. The committee, which Snyder chaired, concluded that neither the Gall-Peters map nor any other rectangular cylindrical projection—plane chart, Mercator's, Lambert's, Behrmann's, Miller's, whatever—was worthwhile. As the first booklet, *Which Map Is Best?* observed, "To force the spherical globe into a rectangle produces extreme shape distortion, but surprisingly most people don't complain." Robinson, who wrote the text, posited an explanation: "To a designer a rectangle is neat: It fits nicely on a rectangular page or wall, and it doesn't leave awkward, empty corners as oval projections do." Even so, he also blamed public familiarity with the rectangular Mercator grid.

Rather than engage Peters's supporters in yet another hissing match, the committee drafted a resolution "strongly urg[ing] book and map publishers, the media, and government agencies to cease using rectangular world maps for general purposes or artistic displays." Severe distortion is inherent in the "straight edges and sharp corners" of any rectangular map, which not only "represent[s] most distances and direct routes incorrectly" but "portray[s] the circular coordinate system as a squared grid." Frequent sightings of a severely distorted world map, the resolution claimed, made it "look right" and contributed to "serious, erroneous conceptions [of] large sections of the world." Borrowing from the Peters playbook, the committee attacked only one map by name—in the resolution's concluding sentence: "The most widely displayed rectangular world map is the Mercator (in fact a navigational diagram devised for nautical charts), but other rectangular world maps proposed as replacements for the Mercator also display a greatly distorted image of the spherical earth." The implication was clear: Mercator's map provides an inappropriate worldview but Peters's projection, though different, was no solution.

Professional cartographers and geographers (or at least the organization leaders who approved resolutions on their behalf) were impressed. In addition to the American Cartographic Association, the antirectangular resolution won endorsements from the American Geographical Society, the Association of American Geographers, the Canadian Cartographic Association, the National Council for Geographic Education, the National Geographic Society, and the Geography and Map Division of the Special Libraries Association. A May 1989 mailing to three hundred news organizations and media officials attracted modest attention. The *Wall Street Journal* ran a short item on its front page, and the *Washington Post* included snapshots of the Mercator and Robinson projections with a brief report in its "Science Notebook" column. Otherwise, the resolution got very little play in the press. Either the map flap was stale news, or the Committee on Map Projections lacked the charisma of Peters's supporters.

Not all academics dismissed Peters's arguments. J. B. Harley (1931–91), one of the world's most respected map historians, saw the Peters controversy as an effort to deny the map's ideological role and ridiculed "the hysteria among leading cartographers at the popularity of the Peters projection." The "real issue," he argued, was a power struggle between Peters, whose "agenda was the empowerment of those nations of the world he felt had suffered a historic cartographic discrimination," and professional cartographers, including most academics, "whose power and 'truth claims' . . . were at stake." In a seminal 1991 essay "Can There Be a Cartographic Ethics?" he charged that cartographers, obsessed with accuracy and willfully ignoring societal consequences, were "still closing ranks." Harley wrote from experience: ACSM had asked him to submit the piece to the *Bulletin* but declined to publish it because his interpretation conflicted with the organization's official position on the Peters map.

Harley echoed Peters's indictment of Eurocentric cartography. "The scientific Renaissance in Europe," he argued, "gave modern cartography coordinate systems, Euclid, scale maps, and accurate measurement, but it also helped to confirm a new myth of Europe's ideological centrality through projections such as those of Mercator." The legacy of a colonial Europe-centered world can be found on all inhabited continents, according to research by Thomas Saarinen, a cognitive geographer who analyzed 3,863 sketch maps by students in forty-nine countries. Although evident on all continents, the myth of European centrality is neither pervasive nor strictly Europe-centered. Participants in Africa and Europe typically drew a north-up map centered around the Greenwich meridian but with Africa clearly in the middle. By contrast, most students in East Asia and Oceania sketched a Sinocentric world, and more than one-fifth of Anglo-American students made their maps Americentric. Surely some of the so-called Eurocentric images merely reflect a natural tendency to position oneself near the center.

Willing to credit Peters with highlighting the problem, neither Harley nor Saarinen was eager to recommend a single solution or put all of the blame on the Mercator projection. John Pickles, another geographic scholar who saw value in the Peters map, questioned the impact of Europe-centered displays. "Cartography shares and reproduces the values of the age," Pickles argued, but the map is more a reflection than a source of power.

Harley, Pickles, and Saarinen have a point. A map can be an ideological statement, as the Peters projection clearly is, and anyone who denies the possibility of an ideological role is clearly wrong. That said, a map's effectiveness as an ideological statement does not make it a reliable device for representing area, shape, or relative importance. Favor one role or attribute, and you're likely to slight others.

John Snyder offered an ironic take on ideology and relative size. At age twenty he developed an equal-area map in the shape of an hourglass (fig. 10.4). I reproduced his map in *How to Lie with Maps* to make the point that area fidelity need not confer shape fidelity. Snyder mentioned the map tongue-in-cheek at the closing paragraph of a letter to Arthur Robinson.

> Enclosed is an *equal-area* pseudocylindrical-type projection I devised around 1946, and drew by hand then. I felt it should remain buried in my notes. Now, it seems the time to arrange a big press conference to unveil it because of the way it shows how so much of the world is choking off the Third World economically. It also draws attention to the center—the Third World areas—and it is *finally* equal-area. Besides Peters, this may be the only equal-area projection ever devised, to go by some published statements. If you will lend your name to its promotion, I'll give you a substantial part of the profits. Or maybe our committee should scrap the pamphlet and promote this instead. Sorry, I have to stop to get into my straitjacket.

Don't misread the sarcasm. A fervently antiracist Quaker convert who once resigned from a church because of the minister's insensitive remarks about African Americans, Snyder resented Peters's exploitation of Third World causes.

I'm continually puzzled by assertions that the Peters projection promotes "fairness to all peoples." What does areal equivalence have to do with population, which varies widely in density throughout the developed and less developed worlds? A better solution, I'm convinced, is the area cartogram, also known as a demographic base map when it distorts map area to portray relative population.

I made one myself once, for a population geography textbook. As shown in fig. 10.5, my map is not without flaws, and it's not particularly original. I borrowed the armadillo graticule from a 1953 cartogram whose authors apparently borrowed it from Erwin Raisz's (1893–1968) clever projection of the world onto a torus. I used 1980 headcounts and tried to draw more readily recognizable geographic caricatures, but only for countries with more than five million people. Thus Africa is set off from Europe and Asia, and Ireland is lumped

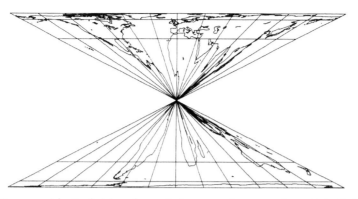

Figure 10.4 John Snyder's hourglass projection, an equal-area map, demonstrates that fidelity of area need not confer fidelity of shape. From Monmonier, *How to Lie with Maps*, 98.

with England, Scotland, and Wales. A larger scale could accommodate additional countries, but not small nations like Djibouti and Luxembourg. Even so, my map not only highlights the numerical prominence of China and India, which look much smaller than Canada on an equal-area projection, but also points out the demographic significance of Indonesia, the world's fourth most populous country but easily overlooked on most maps. For readers curious about smaller

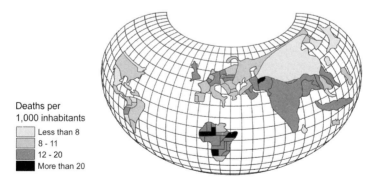

Deaths per
1,000 inhabitants

Less than 8
8 - 11
12 - 20
More than 20

Figure 10.5 World-on-a-Torus population cartogram used as a base map. From Schnell and Monmonier, *Study of Population*, 204.

nations, my coauthor and I included an appendix listing population size and vital rates for all countries, large and small. For precise comparisons, tables of numbers are always more reliable than maps.

A population cartogram can make a strong ideological statement, especially if fairness to all people is more important than fairness to all acres. Even so, demographic base maps don't quite look like maps, and in the international arena they have the added drawback of privileging the more populous nations of Asia over the more numerous but less densely settled countries of Africa. According to historian Jeremy Black, geometric accuracy and even demographic accuracy mattered far less to Third World advocates than the Peters map's distinctive appearance. "Peters struck a chord with a receptive international audience that cared little about cartography, but sought maps to demonstrate the need for a new world order freed from Western conceptions."

Not all Third World supporters agreed. In the early 1980s left-leaning British political scientists Michael Kidron and Ronald Segal used several cartograms in their startlingly innovative *State of the World Atlas.* Only one is based on population; other cartograms use map area to dramatize contrasts in armaments, food supply, and government income. Rather than risk distracting readers with unusual map projections, the authors plotted most of their "angry facts" on the Winkel tripel projection (see the right side of fig. 9.8), a compromise projection adopted two decades later by the comparatively staid National Geographic Society. A note in the corner of the atlas's first map indicates that Kidron and Segal had rejected the Peters map as needlessly distorted and potentially confusing:

> Since the world is virtually spherical it is geometrically impossible to produce a completely accurate world map, on a flat sheet of paper, without some distortion or modification.
>
> The Mercator projection of 1569 and the Peters Projection of 1977 display two extremes of such distortions. However, Winkel's "Tripel" is used throughout this atlas as a familiar and relatively fair, "equal area" projection.

Small insets of the Peters and Mercator maps illustrate these "extremes," while quote marks around "equal area" acknowledge that Oswald Winkel avoided extreme distortions of shape by allowing small distortions of area. Because readers need to focus on the maps' symbols and distribution, a "familiar" projection is an asset.

Peters tested the relative merits of familiarity and visual ideology with his own world atlas, published in 1989. In addition to 246 world maps addressing economic, social, and political topics similar to those in the *State of the World Atlas,* the *Peters Atlas of the World* has a topographic section in which each of 43 two-page maps covers one-sixtieth of the earth's surface at more or less the same scale. In extending the Peters doctrine of areal fairness to relatively detailed region maps, the topographic section grossly overemphasizes sparsely settled areas in Saharan Africa, Greenland, and Antarctica at the expense of India, Indonesia, and other densely populated countries. To avoid extreme stretching on his topographic maps, Peters abandoned his signature projection, secant at 45° N and S, in favor of locally centered cylindrical projections. And in polar zones, where a cylindrical graticule would cause pronounced east–west stretching, an azimuthal projection lets the meridians converge (realistically) to a point. Additional compromise is apparent in the understandable suppression of maps that would cover large expanses of ocean in order to afford equal treatment to the Seychelles, the Cape Verde islands, and other outposts inconveniently distant from a continent. They're present, but only as tiny specks on a very small-scale world map. Fairness suffers further as some places disappear into the gutter between facing pages while others show up on more than one regional map.

Academic cartographers found much to criticize. Many of the color symbols are unnecessarily garish, for instance, and place names are difficult to read on world maps printed eight to a page. Looking beyond design flaws, several reviewers ridiculed the hype-filled promotional material. Russell King and Peter Vujakovic, writing in the British journal *Geography,* validated their title "Peters Atlas: A New Era of Cartography or Publisher's Con-Trick?" by quoting the claim, "This Atlas represents the greatest single advance in map-making in over 400 years."

In the *Bulletin* of the Society of University Cartographers, Vujakovic applauded the "interesting and varied" content of the thematic section but questioned the assertion, in the atlas's two-page introduction, that " 'no interpretation or evaluation of information has been undertaken . . .' in order not to detract from the user forming '. . . an objective and unprejudiced personal picture.' " While interpretative descriptions afford an opportunity for bias, few cartographic decisions are as subjective as the selection of themes and data. Equally deplorable was the atlas's lack of dates and source notes for individual maps.

Peters's more strident claims were toned down in 2002, when the Hammond World Atlas Corporation published the *Hammond Compact Peters World Atlas,* printed in Germany and subtitled *The Earth in True Proportion.* Hammond's 2003 college catalog (fig. 10.6) touts the book as "a distinct alternative and enhancement to all other atlases . . . the first atlas ever to depict countries, continents, regions and their relationships to each other according to true land mass." Insofar as no other world atlas I know of relies exclusively on equal-area projections, the claim seems valid. What's more, the offer of a Peters Combo Pack that includes the more conventional *Hammond New Comparative World Atlas* reflects a new marketing strategy that emphasizes complementarity.

In linking two well-known cartographic brands, Hammond and Peters, the new pitch portrays the Peters atlas as an essential supplement. "Which map projection is 'better,' one that shows true land mass but distorts shape, distance and direction, or vice versa?" Clever phrasing insofar as "vice versa" implies that world maps in the second atlas don't distort shape, distance, or direction. Hype aside, the two volumes are fitting complements because continental maps in the *Comparative* atlas are cast on the innovative "optimal conformal" projection designed by physicist Mitchell Feigenbaum to minimize distortions of large and small shapes as well as distance. To accommodate more reactionary tastes, Hammond's catalog offers a Mercator world map as either a "designer edition" wall map or a page-size laminated version bundled with the company's *Scholastic New Headline World Atlas.*

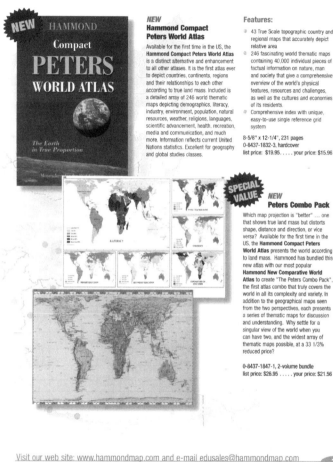
Figure 10.6 The Hammond 2003 college catalog devoted a full page to the *Hammond Compact Peters World Atlas.*

In the 1990s the Peters map became an emblem of diversity awareness, marketed as the cornerstone of a catholic cartography in which
dissimilar images of the world promote cultural sensitivity. An influential advocate is HR Press, which claims "the largest selection of cultural diversity training materials for the workplace." Its Web site offers
the Peters Map Seminar Pack, which includes Peters and Mercator
maps as well as an "Upside Down World Map" formed by casting the
Van der Grinten projection (see the left side of fig. 9.8) with the South
Pole at the top. ODT, Inc., another diversity-awareness publisher, sells
three different south-up maps as well as the conventional Peters projection, the Peters atlas, and *Seeing through Maps,* a book based on the
notion that any map (but particularly a world map) reflects a point of
view. The authors, Ward Kaiser and Denis Wood, discuss the advantages and disadvantages of more than twenty world maps, including
population cartograms, and acknowledge criticisms of Peters's map, including its pronounced stretching near the poles and equator. They re-
ject Peters's claims of priority, but they argue that his projection "has
shaken up cartography and been of enormous value in getting people
to critique and understand the images they are presented with."

ODT's success in marketing the Peters map includes the February
28, 2001, episode of the NBC political drama *The West Wing,* in which
the fictional Organization of Cartographers for Social Equality lobbies
presidential staff to put the Peters map in every public school. Presidential press secretary C. J. Cregg overcomes her initial shock of seeing the strangely stretched continents and becomes a supporter.
Inquiries and sales leaped. Pro-Peters organizations seized the opportunity to promote the Peters message, and on its Friendship Press Web
site the National Council of Churches resurrected a familiar straw
man: "The Mercator was designed for navigation and is still valuable
for that purpose, but it gives a wildly distorted sense of size and position. The Peters shows how large, and where, each country is."

A more recent ODT product suggests that the Peters map, as an
icon of cultural diversity, might be a bit stale. In August 2002 ODT an-
nounced the Hobo-Dyer projection, a cylindrical equal-area map secant at 37° 30′, midway between the standard parallels of the Behr-

mann and Gall-Peters projections and only slightly different from
Trystan Edwards's worldview, secant at 37° 24'. The projection's name
reflects the collaboration of ODT principals *Ho*ward Bronstein and
*Bo*b Abramms with Mick Dyer, a designer at Oxford Cartographers,
the British firm that produced the artwork for the *Hammond Compact
Peters World Atlas*. Particularly striking is the projection's presentation as two maps, printed back to back (fig. 10.7). On one side a conventional north-up view reflects a notable lessening of the north–south
stretching that undermined Peters's portrayal of tropical nations. On
the other side a south-up map centered on 150° E moves Australia
from a peripheral position to center stage. Although south-up maps,
particularly popular in Australia, are not new, the two-sides-of-a-coin
presentation was sufficiently dramatic to capture the endorsement of
Nobel laureate Jimmy Carter. According to a December 6, 2002, ODT
press release, "When President Jimmy Carter receives the Nobel Peace
Prize on December 10th in Oslo, Norway he will take a map developed
and published in Amherst, Massachusetts. The Carter Center chose
ODT's new Hobo-Dyer map to display the 68 countries around the
world in which the Center has worked since 1982." The endorsement

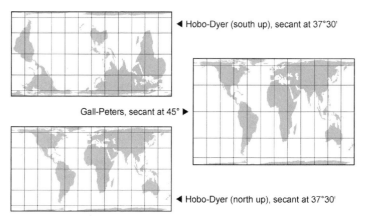

Figure 10.7 A comparison of the Gall-Peters projection with complementary orientations of the Hobo-Dyer projection. Note that the south-up view is centered on 150° E to
place Australia near the center.

Figure 10.8 A world map on the Carter Center Web site decorates a menu for retrieving descriptions, by country, of the organization's humanitarian activities. From http://www .cartercenter.org/ activities/activities.asp?submenu =activities/.

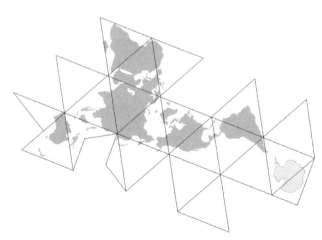

Figure 10.9 Geographers Paul Knox and Sallie Marston used this orientation of R. Buckminster Fuller's innovative "Dymaxion" map in *Places and Regions in Global Context,* an elementary textbook in human geography. For simplicity they removed the triangular framework and Antarctica. The word *Dymaxion* and the horizontal version of the projection are trademarks of the Buckminster Fuller Institute.

seems genuine, but a search of newspaper databases failed to confirm Carter's use of this post-Peters perspective. And three months after his trip to Oslo the Carter Center Web site decorated its activities menu with a world map on what obviously is not an equal-area projection (fig. 10.8).

I compared the Carter map to some well-known rectangular projections but couldn't find a match. Someone apparently compiled an ad hoc world map from diverse sources—perhaps a Mercator map was one of them—and moved Africa and Australia away from Eurasia to make them stand out. Note the gaping Strait of Gibraltar and the partition of New Guinea. Despite these liberties, the map is innocuous. Carter Center officials and their Web designer obviously consider humanitarian good works far more important than relative size, which must be distorted slightly if a world map is not to mangle shape.

If one wants a more eye-catching low-distortion world map, there are numerous options. A good example can be found in a textbook on world region geography by Paul Knox and Sallie Marston, who invoked an innovative map projection developed a half-century earlier by famed architect-inventor R. Buckminster Fuller. To make distortions of area and angles essentially invisible, Fuller mapped the globe onto an icosahedron, a three-dimensional solid consisting of twenty equilateral triangles (fig. 10.9). Yes, there's distortion, but because the triangles nicely mimic small sections of the globe's surface, stretching is largely negligible on small-scale maps. The Buckminster Fuller Institute uses a version with the main band of triangles aligned horizontally as its trademark. By contrast, Knox and Marston rotated the image clockwise about twenty-one degrees, stripped away the triangular framework, and zapped Antarctica, which contributed nothing to their demographic and economic maps. The awkward orientations of Australia and the Americas underscore the trade-offs inherent in map projection.

Points of View

Arno Peters made it difficult to view the Mercator projection as merely another navigation tool. Overseas conquests by the Spanish, the Portuguese, and the Dutch, to name a few, depended on sea power, and a map that helped colonial navies reach distant shores seems at first glance a worthy scapegoat for European exploits in Africa and Asia. Equally problematic are questions raised by the projection's more general uses. Are its inappropriate adoption for nineteenth-century reference atlases and its persistence on post–World War II wall maps merely reflections of its role in navigation, or might sinister political motives be at work?

Maybe, but probably not. The Mercator projection's formidable societal momentum demands multiple explanations, which include collective memory, brand-name recognition, and institutional inertia. Both the map and its author are well known, to be sure, and their coastlines look right to many people. Commercial mapmakers are understandably reluctant to snub buyers eager for a time-tested product, and designers seeking a traditional tone can readily exploit its familiar authority. Cartographic educators have railed against its misuse for at least a century but cannot stamp out ill-advised Mercator world-views. And for all their pleas for fairness, neither can Peters and his disciples.

Despite its cachet, the Mercator projection is greatly overrated as a symbol of Western imperialism. Functional maps were essential for efficient navigation but not uniquely so. Straight rhumb lines helped naval commanders, merchant captains, and slave traders go about their business, but so did the caulking for their hulls, the timber for their masts, and the canvas for their sails. Equally questionable is the Mercator map's influence on social thought and world politics. Did Europe's rulers and merchants need wall maps or world atlases to justify their actions? Did maps that inflated the size of the British Empire stifle whatever remorse nineteenth-century Britons might have had about racism and economic slavery in Africa or India? More to the point, did anyone ever die because of the Mercator projection?

I feel like a heretic to say it, but cartographic scholars engrossed in ideology and empowerment have vastly inflated the importance of maps, and with it the significance of their scholarship. While maps can be influential in contemporary disputes over boundaries or environmental impact, broader geopolitical impacts are more difficult to gauge. Seduced by the "power of maps" as an intellectual agenda, self-proclaimed theorists demand little evidence for innovative mono-causal arguments (like Peters's) that might seem sensible were their proponents not aggressively trouncing equally plausible explanations.

A case in point is Brian Harley's endorsement of Arno Peters's effort to promote "fairness" to Third World nations without questioning how the Peters projection, or any other map for that matter, might achieve this worthy goal. While Harley merits praise for his eloquent analysis of "cartographic silences," a concept based on maps' ability to manipulate opinion by omitting or suppressing information, his willingness to excuse the "silences" of Peters's proponents, who blatantly ignore existing equal-area maps, is puzzling.

Need for verification doesn't stop there. In questioning the motives of Peters's critics, Harley overlooked an opportunity to challenge academic cartography's untested assumption that the Peters map can seriously impair public understanding of geography. Although superior projections abound, the evils of the Peters map are easily exaggerated. Do its users really think Africa looks that way? Do they never look at a globe, or at other maps? Are map users complete idiots?

An abundance of dysfunctional designs in news publications and academic journals suggests that maps as a whole are remarkably robust. By this I mean that a map need not be well designed or user friendly to be informative, at least among conscientious users who know their geography and read maps carefully. We live with distorted maps—we have no choice—and a map that clobbers the shapes or sizes of continents is not intellectually poisonous, especially for users acquainted with other frameworks. For these viewers, there is ample room for both the Peters and the Mercator. The real problem is broader ignorance of maps, geography, and geometry. Among a map-savvy public, the Peters projection would have few adherents.

A persistent concern is the Mercator map's areal distortion, especially troublesome when uniform dots representing a fixed number of people, hogs, or apple trees portray variations in density. As every mapmaker knows, or should know, an equal-area projection is essential if a dot-distribution map is to reveal reliable contrasts between high-density and low-density regions. For other types of map the cartographer must weigh distortion of area against distortions of angles, distances, and directions. Also relevant are the gross shapes of continents, faithfully represented only on a globe. As Goode's equal-area map (see fig. 9.5) demonstrates, interruptions over water can minimize shape distortion by allowing each region its own, locally centered map projection. As Robinson's compromise projection (see fig. 9.7) shows, modest concessions to area distortion toward the poles can foster relatively realistic continents without ripping apart the oceans. And when true angles are important on meteorological maps covering tropical and temperate latitudes, the conformal cylindrical projection pioneered by Gerard Mercator is peculiarly appropriate.

Although Peters's supporters treat areal distortion like a crime against humanity, modest distortions of relative area can be highly useful on compromise projections like Robinson's and Winkel's tripel (see the right side of fig. 9.8). Truth be told, exact area equivalence is often wasted on world maps intended to foster comparisons of relative size. Precise visual comparison is impossible because map generalization precludes an exact portrayal of coastlines and national boundaries, and because irregular shapes interfere with perception of shape. If you don't believe me, show a friend an equal-area map of the United States and ask whether Florida is larger than Georgia. The Sunshine State's panhandle and greater north–south extent trick most people into thinking, incorrectly, that it's bigger than its more compact neighbor. A similar L-shaped outline suggests that Africa on an equal-area map might look slightly larger than it should.

If any projection is worth denouncing, it's the vaguely named "geographical projection" popular among users of geographic information system (GIS) software. Used occasionally for maps of the conterminous United States (fig. 11.1), it's a nonprojection that treats longitude and latitude as rectangular (x, y) coordinates—an undisguised throwback to the plane chart of pre-Mercator sailing. Because meridians cannot converge, the map inflates the areas of Montana, North Dakota, and other northern states.

NPIAS Airports

Figure 11.1 GIS software encourages use of the "geographical projection," which enlarges the relative area of northern states by using longitude and latitude as rectangular coordinates. From U.S. Department of Transportation, Federal Aviation Administration, *National Plan of Integrated Airport Systems,* appendix B.

Peters's complaint that the Mercator projection favors northern countries at the expense of the Third World finds favor among post-colonial deconstructionists poised to slay dead dragons. Critical theorists suspicious of government mapmaking see topographic and other maps as biased representations, subverted to whatever questionable agenda "the state" might promote. Although a potential for bias exists, broad assumptions of conscious or subliminal malevolence trivialize commonsense notions of bias and agenda. In my experience, the bias of ignorance, the bias of sloppiness, and the bias of tradition, individually or collectively, are far more prevalent than the bias of political ideology.

Mistrust of the Mercator map is not new, as historian Susan Schulten discovered in her examination of American geography textbooks from the 1940s. Controversy over the centering of world maps before and after the Japanese bombed Pearl Harbor is particularly revealing. While some geopolitical strategists attacked the traditional Greenwich-centered world for downplaying the strategic significance of the Pacific Ocean, others condemned world maps centered on 90° W (roughly the longitude of Chicago) for promoting a sense of "psychological isolation" among Americans by showing the Atlantic and Pacific Oceans as "a sort of 'Maginot Line' around the Western Hemisphere." Postwar anxiety about a possible over-the-pole attack from the Soviet Union led to "replac[ement of] the massive ocean buffers of the Atlantic and the Pacific on the Mercator projection with the relatively insubstantial Arctic, a 'new mediterranean' surrounded by the two political superpowers." Although Schulten mentions the Mercator projection by name, she's more concerned with how maps were centered than with their treatment of relative area. In this sense "Mercator" refers more generally to any equatorially centered cylindrical map projection in much the same way "Coke" sometimes means Pepsi.

The most trumped-up charge against the Mercator map is its alleged Eurocentrism. Horizontally, any projection centered on the Greenwich meridian is also centered on Africa. Vertically, any cylindrical projection on which the poles lie an infinite distance from the equator can be configured with Europe at the center. Insist on show-

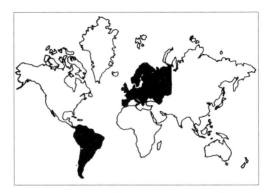

Figure 11.2 A pro-Peters rendering of the Mercator projection, framed as a Europe-centered straw man. From Kaiser and Wood, *Seeing through Maps*, 7.

ing all of Greenland and not a hint of Antarctica, and a Mercator map becomes Eurocentric (fig. 11.2). Whatever social or political consequences this configuration might have, the only examples I've seen are straw man renderings designed to promote the Peters map.

Repeat an inaccurate statement assertively, and uncritical readers accept it as fact. This aphorism might explain Simon Winchester's catty complaint about fellow British writer Nicholas Crane's recent biography of Gerard Mercator. In a review for the *New York Times*, Winchester chides Crane for "not linger[ing] on the social and political implications of Mercator's map" and not "tr[ying] to undo the wrongs that Mercator perpetrated more than four centuries ago." In addition to famously inflating northern lands, the great mapmaker "chose to set his equator two-thirds of the way down his sheet, the better to give his native Flanders a more suitably dignified position on the chart." Had that been Mercator's motive, I doubt that Crane would have missed it. True, the 1569 map's equator is a bit below the map's center, but a look at the original chart (or the tracing in fig. 4.1) indicates that "two-thirds of the way down" is an exaggeration.

Empower yourself by confronting the myth of Mercator's Eurocentrism. The next time you see a Mercator map—assuming it's not there just to prove Peters's point—take a tape measure or piece of string and find its real center. What you find will probably not be greatly different from the Mercator wall map hanging decoratively outside the map room at the Syracuse University's Bird Library. Bisected vertically by the 33rd parallel, the map is clearly centered on North Africa. Yes, its equator is pushed southward, more so than on Gerard Mercator's original, but Europe is definitely not at the center. I have no idea who purchased the map or when it arrived. Rand McNally copyrighted it in 1993, and its faded colors indicate a decade's exposure to light. While another Mercator map would not be my choice for a replacement, the one that's there is harming no one.

Peters's attack is neither the first nor the most serious threat to the Mercator projection. As noted in chapter 6, replacement of the Mercator on aeronautical charts by the Lambert conformal conic attests to the diminished importance of rhumb lines. John Q. Stewart highlighted the latter's advantages in a 1943 *Geographical Review* essay: "The civilian flyer over a country well mapped and abounding in easy navigational marks (rivers, large bridges, railroads, race tracks) and provided with radio aids and lighted airways does not trouble to engage in pin-point plotting. For him the relative constancy of scale of the Lambert outweighs everything else; and it is important to give him what he needs." That year Consolidated Vultee Aircraft Corporation touted another air-age tool, the azimuthal equidistant projection, in a booklet on maps and air supremacy: "[A]lthough a Mercator map of the world shows Greenland many times its actual size, it is nevertheless ideal for surface navigation because it shows true compass directions. And . . . the Azimuthal Equidistant projection, while it may show some areas sadly out of shape, provides an ideal chart on which the aviator can plot a true course and measure his distance from a given point." Nowadays plotting the route is less important than entering the right coordinates into an electronic system that uses satellite positioning, radar, and radio beacons to stay on course. Computers that can fly the plane and display any map the pilot wants have redefined the role of the paper aviation chart. And similar electronic aids for mariners have diminished the significance of the Mercator sailing chart.

Despite these inroads, Mercator's map still decorates walls and page layouts. On commercial wall maps in settings that invite close inspection, the projection's conformality is an asset for viewers concerned more with local geography than global comparisons. Only when the viewer steps back from the display does area exaggeration intrude. As artistic flourishes, small-scale Mercator maps bear little trace of their Cold War role in John Birch Society propaganda, designed to inflate the size, and hence the threat, of China and Russia. Despite the potential for more contemporary political protest, I have yet to see one used in spite to denounce pro-Peters political correctness. As far as I can judge, artists who invoke the Mercator worldview are merely exploiting the visual equivalent of a classic style like the penny loafer. Surely the *Atlantic Monthly,* in laying out a Mercator framework for The World in Numbers features examining global issues (fig. 11.3), is not thumbing its nose at the Third World. Nor is the

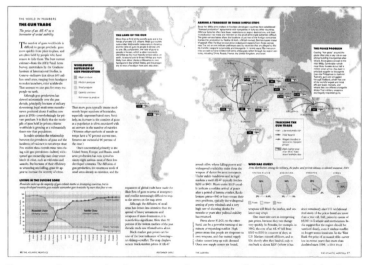

Figure 11.3 Demonstrating that maps can decorate as well as inform, the *Atlantic Monthly* embeds a Mercator projection in the standard two-page layout used when its The World in Numbers feature addresses global patterns. From Peck, "Gun Trade."

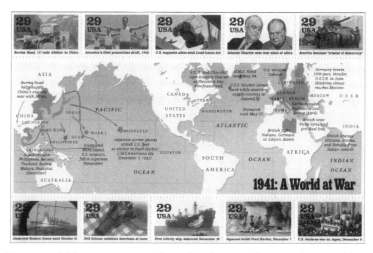

Figure 11.4 A Mercator map centered on 90° W dominates the stamp sheet issued by the U.S. Post Office in September 1991 to commemorate the fiftieth anniversary of World War II.

U.S. Postal Service, in placing a Mercator map at the center of a stamp sheet commemorating American involvement in World War II (fig. 11.4), using artistic ambiance as an excuse for geopolitical propaganda. As in other formal displays, Mercator's rectangular framework aligns nicely with the stamp sheet's edges and perforations.

Like any popular icon under attack, the Mercator map has its defenders. Among its earliest and more moderate supporters were Charles Deetz and Oscar Adams, the leading federal experts on map projection during the 1920s and 1930s. Their 1921 treatise *Elements of Map Projection* interprets exaggerated areas in higher latitudes as an asset because "in the consideration of the various evils of world maps, the polar regions are, after all, the best places to put the maximum distortion." Favoring rectangular frameworks over curved parallels and converging meridians, Deetz and Adams denounced "charts having correct areas with cardinal directions running every possible way [as] undesirable." Articulating an opinion rarely heard these days, the pair maintained that "the Mercator projection not only is a fixture for nautical charts, but . . . plays a definite part in giving us a continuous conformal mapping of the world."

Later defenders were more strident. In a 1943 *Harper's Magazine* article titled "Those Misleading New Maps," University of Minnesota astronomy professor Willem Luyten declared the Mercator chart "the standard map of the world to date" and asserted that "shape is much more important than size." In his view, azimuthal air-age perspectives were unnecessary—the real problem was cartographic ignorance: "All of us who know our geography and are teaching it have known right along what the errors and the shortcomings of the Mercator map are. And so have the navigators of ships and planes who use them. But not so the uninformed amateurs of the global geography, who appear to have accepted the Mercator map as gospel truth: they were Flat-Earthers. And now all of a sudden they have discovered that the earth is round." Readers eager to follow the war and prepare for a wider use of aviation were told to get a detailed Mercator map and a globe.

While Luyten comes across as a fuddy-duddy, Hoover Institution economist Thomas Sowell personifies conservative resistance to leftist "moral preening." His 1995 book *The Vision of the Anointed,* which emphatically rejects the Mercator map's alleged Eurocentrism and Western arrogance, is less caustic than his 1990 *Washington Times* op-ed, which paints a sarcastic picture of political correctness in which "deep thinkers find [Mercator's map] sinister, if not racist." Rhetoric aside, Sowell's appreciation for the Mercator map is more personal than ideological. He had an early interest in map projections and found a Mercator wall map a convenient prop for a hastily improvised but well-received speech in a ninth-grade English class. "Don't mess with the Mercator projection," he warned. "It rescued me when I needed it, and now it's time to return the favor."

In addition to exposing the ideological roles of map projections, the Peters controversy revealed once again the ignorance of map projection, and geometry in general, among the public, the media, and even some academic geographers. Anyone who teaches geography or craves an appreciation of the environment and world affairs needs a basic understanding of how map projections work, how they can be manipulated, and how narratives of their misuse and alleged misuse are readily distorted. Rather than banish the Mercator map, an enlightened society needs to appreciate the limitations of globes and the marvelous flexibility of map projection. With "virtual globes" and inexpensive, highly interactive cartographic multimedia so widely available, mapmakers and teachers have little excuse for inappropriate choices and uninspired pedagogy.

Notes

CHAPTER ONE: BEARINGS STRAIGHT — AN INTRODUCTION

General Sources

Although this introduction draws on numerous sources cited for later chapters, a few general works were especially helpful. My prime source for mathematical and general historical facts about map projections is Snyder, *Flattening the Earth.* For detailed descriptions of Mercator's maps and globes, I relied heavily on Karrow, *Mapmakers.* English translations of Latin inscriptions on Mercator's world map of 1569 are from "Text and Translation of the Legends . . . by Gerhard Mercator."

Notes

2   . . . Willebrord Snell (1580–1626) coined . . . : See Rickey and Tuchinsky, "Application of Geography to Mathematics."

4   According to cartographic historian . . . : Karrow, *Mapmakers,* 382–83.

6   A typical portolan chart . . . : Generously defined, the Mediterranean region extends roughly 2,500 miles E–W and 1,200 miles N–S, for a total of 3 million square miles. Applying the Earth's larger, equatorial diameter of 7,926 miles to a sphere yields an area of roughly 200 million square miles, of which 3 million square miles would be 1.5 percent; given these calculations, "less than one-fiftieth" (2 percent) is a safe estimate.

8   Some scholars think Mercator . . . : For the approximation hypothesis, see Nordenskiöld, *Facsimile Atlas,* 96. For the graphic hypothesis, see McKinney, "The Wright Projection."

8   And in 1845, Bond suggested . . . : See Maor, *Trigonometric Delights,*
     178.

9   "a high rocky island" . . . : Wallis and Robinson, *Cartographical Innovations,*
     188.

10   Latitude could be figured . . . : For a fascinating account of Harrison's strug-
     gle, see Sobel, *Longitude.*

11   According to Robert Karrow, nineteen copies . . . : Karrow, *Mapmakers,* 389,
     392.

11   An instrument maker and physician . . . : Schnelbögl, "Life and Work of . . .
     Erhard Etzlaub."

13   Englisch argues . . . : Englisch, "Erhard Etzlaub's Projection," 115.

13   "uses a Mercator projection . . ." : See Hellemans and Bunch, *The Timetables
     of Science,* 70.

13   As a key sentence in its caption . . . : Stephenson, "Chinese and Korean Star
     Maps and Catalogs," 534.

13   "the principle remained obscure . . ." : Snyder, *Flattening the Earth,* 48.

13   "might well have been termed . . ." : Deetz and Adams, *Elements of Map Pro-
     jection,* 5th ed., 173.

14   Noteworthy adoptions . . . : For discussion of seventeenth- and eighteenth-
     century adoptions of Mercator's projection, see Snyder, *Flattening the Earth,*
     48, 59–60.

14   In 1919 Vilhelm Bjerknes . . . : For discussion of meteorological uses of the
     Mercator projection, see Monmonier, *Air Apparent,* 218.

15   "Let none dare to attribute . . ." : Deetz and Adams, *Elements of Map Projec-
     tion,* 5th ed., 104.

Chapter Two: Early Sailing Charts

General Sources

The most comprehensive recent treatment of portolan charts is Campbell, "Portolan Charts." I also drew on Bagrow, *History of Cartography,* esp. 61–66; Lanman, *On the Origin of Portolan Charts*; Mollat, *Sea Charts of the Early Explorers*;
Snyder, *Flattening the Earth*; Stevenson, *Portolan Charts*; Taylor, *Haven-Finding
Art*; and Whitfield, *Charting of the Oceans.*

Notes

18   In their handbook . . . : Wallis and Robinson, *Cartographical Innovations,*
     12–17.

19   Figure 2.3: In tracing this figure, I deliberately widened the rhumb lines that
     delineate the square grids and straightened lines warped by the shrinkage
     or swelling of the vellum.

20   Treatment included soaking . . . : For a fascinating introduction to vellum making, see Reed, *Nature and Making of Parchment,* esp. 75 – 96. As Reed notes, there is no obvious difference between vellum and parchment.

20   . . . lists forty-six individuals . . . : Campbell, "Portolan Charts," 449 – 58.

21   In 1354 King Peter . . . : Ibid., 440.

21   The earliest surviving record . . . : Ibid., 439.

22   "the thousands of ordinary charts . . ." : Ibid., 446.

22   Some chartmakers apparently . . . : Ibid., 391.

23   "drawn on a square grid" : Lanman, *On the Origin of Portolan Charts,* 53.

23   . . . Waldo Tobler, a pioneer . . . : Tobler, "Medieval Distortions."

23   And in a cartometric analysis . . . : Loomer, "Cartometric Analysis." Also see Loomer, "Mathematical Analysis of Medieval Sea Charts."

24   Some historians recognize . . . : Taylor, *Haven-Finding Art,* 111.

25   "It was several times reported . . ." : Whitfield, *Charting of the Oceans,* 29.

28   Devised around AD 100 . . . : Dilke, *Greek and Roman Maps,* 72–79.

28   In the late fifteenth century . . . : Campbell, "Portolan Charts," 385.

29   As this latter audience . . . : Snyder, *Flattening the Earth,* 6 – 8; and Whitfield, *Charting of the Oceans,* 40, 44. For a facsimile of Waldseemüller's *Carta Marina,* see Whitfield, *Image of the World,* 54 – 55. For a facsimile of Ribero's *Carta Universal,* see Nebenzahl, *Atlas of Columbus,* 94 – 95.

Chapter Three: Mercator's Résumé

General Sources

For facts and interpretations of Mercator's life I relied heavily on Karrow, *Mapmakers,* esp. 376 – 406; and Osley, *Mercator.* Karrow, who is curator of maps at the Newberry Library, in Chicago, is also my key source for information on Mercator's cartographic works.

Sources for this chapter also include Averdunk and Müller-Reinhard, "Gerard Mercator"; and Calcoen and others, *Le cartographie Gerard Mercator.* The latter contains an outstanding collection of facsimile excerpts.

Notes

31   Although biographers lament . . . : For a collection of Mercator's correspondence, see van Durme, *Correspondance Mercatorienne.* According to cartographic historian Eva Taylor, the 217 letters were assembled from a variety of scattered sources, focus on discussions of family or religion, and offer no new insights on Mercator's world map of 1569; see Taylor, "Correspondance Mercatorienne [review]."

32   Mercator's first biographer . . . : For an English translation of Ghim's biography, "Vita Mercatoris," see Osley, *Mercator,* 185 – 94.

32 "remarkable and distinguished man" . . . "mild character and honest way of life" : Ghim, in Osley, *Mercator,* 188, 190.

32 "the energetic priest of that city" : Ibid., 185.

32 "82 years, 37 weeks . . ." : Ibid., 194.

32 He considered italic . . . : Osley provides a facsimile of Mercator's pamphlet as well as an English translation; see Osley, *Mercator,* 121–74.

33 Although Gerhard Cremer . . . : Robert Karrow identifies two additional variants: Gerardo Rupelimontano and Gheert Scellekens; see Karrow, *Mapmakers,* 376.

34 His confidants included . . . : Ibid., 407–9.

34 "emerged with strong . . ." : Osley, *Mercator,* 20.

34 "a superb engraver . . ." : Ibid., 21.

34 In 1536 he engraved . . . : Typically, a molded ball of paper, paste, glue, and shellac was built up in layers and reinforced with hemp fibers. For a concise introduction to medieval globe making, see Dekker, *Globes at Greenwich,* 21–23.

35 "commercial success" : Karrow, *Mapmakers,* 378.

35 Figure 3.2: For a facsimile of Mercator's map, which also employed a ten-degree graticule, see Osley, *Mercator,* 66.

35 As close examination . . . : Brown, *Story of Maps,* 159.

36 Aware of the uncertainty . . . : As Nicholas Crane observed, Mercator depicted known coastlines with a continuous line supplemented on its seaward side with short, thin, closely spaced horizontal lines, called *hachures.* By contrast, he represented unknown coastlines with hachures alone. See Crane, *Mercator,* 98; and Nordenskiöld, *Facsimile Atlas,* 91, fig. 54.

36 A key skeptic is . . . : For discussion of Kirmse's findings, see Karrow, *Mapmakers,* 380. For the original article, see Kirmse, "Die grosse Flandernkarte Gerhard Mercators."

37 Although protests . . . : Accounts of the executions differ; my key source is Karrow, commentary accompanying Mercator, *Atlas.*

37 Some writers consider . . . : For an early insight on the issue, see Hall, "Gerard Mercator," esp. 163, 185–86. Hall notes briefly the debate over whether the famed mapmaker was a Flemish Netherlander (and thus likely to be a Catholic) or a German (more likely to be a Protestant). Although agreeing with Belgian historian Jan Van Raemdonck, who considered Mercator a Catholic to the end, Hall mentions a letter from German historian, Arthur Breusing, who thought otherwise.

37 "I find it impossible to tell" : Westfall, "Mercator, Gerardus."

37 "attracted more praise . . ." : Ghim, "Vita Mercatoris," in Osley, *Mercator,* 187.

37 According to . . . : Barber, "British Isles."

38   "a distinguished friend . . ." : Ghim, "Vita Mercatoris," in Osley, *Mercator,* 187.

38   "find the friend" : Barber, "British Isles," 55.

39   Seven subsequent editions . . . : Karrow, *Mapmakers,* 396–98.

39   While working on Ptolemy's . . . : LeGear, "Mercator's Atlas of 1595."

40   Figure 3.3: Arthur Osley, who also used this engraving as a frontispiece, attributes it to Frans Hogenberg; see Osley, *Mercator,* 4.

41   Like other mapmakers, Mercator relied . . . : See Karrow, commentary accompanying Mercator, *Atlas,* 20.

41   The late Clara LeGear . . . : Ibid.

42   "had drawn up . . ." . . . "had sold a large quantity . . ." : Ghim, "Vita Mercatoris," in Osley, *Mercator,* 188.

42   According to map historian . . . : Akerman, "Atlas."

42   "remarkable that . . ." : Ibid., 26.

43   Figure 3.4: Substantially reduced from the color original.

45   In 2000 Octavo Digital Editions . . . : For a review, see Turley, "Atlas Rebound." Octavo is a digital publisher specializing in inexpensive electronic reprints of rare books.

Chapter Four: Revealing Replicas

General Sources

Much of this chapter is based on a 1961 facsimile of Mercator's world map and an accompanying sixty-nine-page introduction by Bert van 't Hoff, published as "Gerard Mercator's Map of the World (1569) in the Form of an Atlas in the Maritiem Museum 'Prins Hendrik' at Rotterdam." Other useful descriptions of the 1569 map include Fite and Freeman, *Book of Old Maps,* 77–80; and Nebenzahl, *Maps from the Age of Discovery,* 126–29. For information about various facsimiles, I relied heavily on Karrow, *Mapmakers.*

Notes

47   When reduced to a black-and-white . . . : See, for example, Bagrow, *History of Cartography,* pl. LXX.

48   Similar renderings . . . : Shalowitz, "Chart That Made Navigation History"; and Chamberlin, *Round Earth on Flat Paper.*

49   The oldest is part . . . : My principal source for information on facsimile reproductions is Karrow, *Mapmakers,* esp. 389. Additional information about Jomard is from French and others, *Tooley's Dictionary of Mapmakers,* E-J volume, 451–52; and Godlewska, "Jomard," 109–35. Unlike later facsimiles of Mercator's 1569 world map, the image for Jomard's version, reproduced lithographically, was transferred to the printing plates mechanically rather than photographically; see Pinson, "Repressed Mimesis."

49   The municipal library . . . : For a description, see Heyer, "Drei Mercator-Karten."

49   Destroyed during World War II . . . : "Text and Translation of the Legends . . . by Gerhard Mercator."

49   A fourth copy . . . : The folded-over facsimile of the Basel copy is tipped in between pages 104 and 105 of Taylor, "John Dee." Dimensions refer to the image; with nearly one-inch margins on all sides, the paper is a bit larger.

51   "the size of the maps . . ." : "Gerard Mercator's Map of the World," 22.

52   "the first printed sea-atlas" : Bagrow, *History of Cartography,* 119. For a description, see van Nouhuys, "Mercator's World Atlas."

52   The evidence is a bound . . . : For discussion of the atlas and its discovery, see Karrow, *Mapmakers,* 393–95; Scott and Goss, "Important Mercator 'Discovery'"; and Varekamp, "Discovery of Manuscript Maps."

53   "the most important . . ." : Karrow, *Mapmakers,* 393.

53   But not any longer . . . : *The Mercator Atlas of Europe: Facsimile of the Maps by Gerardus Mercator Contained in the Atlas of Europe, circa 1570–1572,* ed. Marcel Watelet (Pleasant Hill, Oreg.: Walking Tree Press, 1998). For a concise account of the atlas's purchase, see Reynolds, "First Maps of Britain."

53   If you crave fine details . . . : The scan resolution is approximately 60 pixels per inch—crude by current standards. I estimated the resolution from the image heights of the three leftmost plates, which are 1002, 1002, and 997 pixels tall, for a total height of 3001 pixels. Dividing by the map's reported height of 49 inches yields 61.2, which rounds to 60.

55   "the imperfection . . ." and "can be explained . . ." : Nordenskiöld, *Facsimile Atlas,* 96.

55   In announcing . . . : "Mercator's World Map of 1569."

55   "this beautiful and remarkable . . ." : Hoff, introduction to "Gerard Mercator's Map," 28.

56   Convinced that Etzlaub's maps . . . : Also see Wedemeyer, letter dated August 18, 1931; and more recently, Kretschmer, "Kartenprojektion," esp. 380.

57   The series stimulated . . . : Krücken, "Wissenschaftsgeschichtliche."

57   "rather unlikely that . . ." : Ingrid Kretschmer, e-mail communication, May 15, 2002.

57   . . . the work of Pedro Nuñes . . . : For a concise account of Nuñes's life and work, see López de Azcona, "Nuñes Salaciense, Pedro." Also see Waters, *Art of Navigation,* 71–72.

58   Figure 4.5: Kretschmer attributes this facsimile to De Smet, *Les sphères terrestre.*

58   "with the aid of metallic . . ." : Keuning, "History of Geographical Map Projections," 16.

58   I'm surprised . . . : Ibid., 18.

59   "By following his curved rhumb lines . . ." : Karrow, *Mapmakers,* 389.

59   "that the requirement . . ." : Warntz and Wolff, *Breakthroughs in Geography,*
      69.

59   According to map historian . . . : Meurer's work is cited by Frank Koks, who
      prepared the "Ortelius Atlas" page for the U.S. Library of Congress's Ameri-
      can Memory Web site, http://memory.loc.gov/ammem/gmdhtml/
      gnrlort.html; also see Meurer, *Fontes cartographici Orteliani,* 10–11.

59   Features and place names . . . : Hoff, introduction to "Gerard Mercator's
      Map," 33.

60   Another paradox . . . : Karrow, *Mapmakers,* 389–91.

61   Hondius's map predates . . . : A 1608 edition of Hondius's map was believed
      to be a later revision of a map Hondius first published in 1598; see Hea-
      wood, "Hondius and His Newly-Found Map," esp. 181–82; and Hooker,
      "New Light." In 1993 map historian Gunter Schilder uncovered a copy of the
      missing 1598 Hondius map; see Cohen and Augustyn, "Newly Discovered
      Hondius Map."

CHAPTER FIVE: THE WRIGHT APPROACH

General Sources

My key source on Edward Wright is Parsons and Morris, "Edward Wright." I con-
sulted the *Hydrographic Review,* selected facsimile reprint of the 1599 edition of
Wright's *Certaine Errors in Navigation,* which includes the cover, preface, and
abridged table of meridional parts. Crucial references for Thomas Harriot are
Pepper, "Harriot's Calculation"; and Taylor and Sadler, "Doctrine of Nauticall Tri-
angles." For commentary on Henry Bond's discovery, I relied largely on Cajori,
"On an Integration"; Maor, *Trigonometric Delights,* 165–80; and Rickey and
Tuchinsky, "Application of Geography to Mathematics." My principal sources for
J. H. Lambert are his *Notes and Comments*; and Maurer, "Johann Heinrich Lam-
bert." Incisive treatments of Tissot's indicatrix include Laskowski, "Traditional
and Modern Look"; Marschner, "Structural Properties"; and Snyder and Voxland,
Album of Map Projections.

Notes

65   "not at this time . . ." : Wright, "Preface to the Reader," in *Certaine Errors*
      (1599), *Hydrographic Review* facsimile reprint, 94.

65   . . . the complete table . . . : See Wright, *Certaine Errors in Navigation,* 3rd ed.
      (1657), 14–36.

69   No less apparent . . . : For a concise discussion of the Christian-Knight map,
      see Hind, *Engraving in England,* 1:176–77; and Shirley, *Mapping of the
      World,* 218–19.

69   "I hear that you are . . ." and all other quotations in this paragraph, except
      Wroth's, are from Wright's "Preface to the Reader," in *Certaine Errors* (1599),
      Hydrographic Review facsimile reprint, 90.

69   "the most inept . . .": Wroth, *Way of a Ship*, 66.

70   "a friend of mine . . .": Quoted in Parsons, "Edward Wright," 62.

70   According to maritime historian . . .: Waters, *Art of Navigation,* 220–21.

71   Hungry for accurate information . . .: Davis, *Voyages and Works,* 2:4.

72   "adverse external circumstances . . .": Lohne, "Harriot (or Hariot), Thomas,"
      128.

73   Halley titled his article . . .: Halley, "Early Demonstration," 202.

73   " . . . not without a long train . . ." and "Wherefore having attained . . .": Ibid.,
      203–4.

74   "his strange appearance . . .": Lohne, "Lambert, Johann Heinrich," 7:597.

75   I suspect, though, . . .: Maurer, "Johann Heinrich Lambert," 72.

Chapter Six: Travelers' Aide

General Sources

For information on the development of navigation instruments I relied on Taylor
and Richey, *Geometrical Seaman.* My principal source for mid-twentieth-century
thinking on map projections for navigation was Stewart and Pierce, *Marine and
Air Navigation.* Key sources for my discussion of Maury were Lewis, *Matthew
Fontaine Maury*; Weber, *Hydrographic Office,* esp. 16–30; and Williams, *Matthew
Fontaine Maury.* For information on Ferdinand Hassler and his development of
the polyconic projection, I relied largely on Cajori, *Chequered Career*; Schott,
"Comparison of the Relative Value"; and Stachurski, "History of American Pro-
jections." Basic sources for information on aeronautical charts include Bryan,
"Aeronautical Charts"; and Ristow, *Aviation Cartography.* Information on meteo-
rological cartography is drawn largely from Gregg and Tannehill, "International
Standard Projections"; and Griggs, "Background and Development."

Notes

80   "THERE is now going . . .": Haselden, *Reply to Mr. Wilson's Answer,* 14.

80   "[S]ince 'tis very reasonable . . .": Haselden, *Description and Use,* i–ii.

81   " . . . the Mercator's-Chart . . .": Ibid., ii.

81   "I cannot but think . . .": Ibid., viii.

81   "many groundless Objections": Ibid., 27.

81   As modern textbooks . . .: For example, "Mercator charts have the special ad-
      vantage that a distance as great as 20° or 30°—1200–1800 nautical miles—
      along any rhumb line can be very approximately measured as a whole instead
      of summed by parts." See Stewart and Pierce, *Marine and Air Navigation,* 33.

84   Between 1848 and 1861 . . . : Weber, *Hydrographic Office,* 18.

85   . . . Swiss-born mathematician-geodesist . . . : Authorized by Congress in
      1807, the Survey of the Coast had a fitful early history, disrupted by politics,
      uneven funding, and war with Britain. Hassler was first hired in 1811 to de-
      vise procedures and buy equipment. Appointed superintendent in 1816, he
      left two years later when Congress called for staffing by military personnel.
      His 1832 appointment was thus a reappointment. See Weber, *Coast and
      Geodetic Survey,* 1–5.

85   "You come to 'spect . . ." : "Coast Survey," 507.

90   "This distribution of the projection . . ." : Hassler, "Papers on Various Sub-
      jects," 407–8.

90   The Coast Survey's early charts . . . : For insights on the emergence of the
      polyconic projection as the official cartographic framework of the Coast Sur-
      vey, see Schott, "Comparison of the Relative Value." For information on
      Bonne's projection, see Snyder, *Flattening the Earth,* 60–62.

90   In 1910, after years . . . : After the navy "urgently requested" conversion to
      the Mercator projection, the Survey appointed a board to study the request
      and make a recommendation. The board recommended conversion, and the
      Survey devised a plan to eliminate the "old-style charts." See U.S. Coast and
      Geodetic Survey, *Report of the Superintendent,* 11. According to Aaron
      Shalowitz, who downplayed the navy's role, "The real impetus to general
      conversion to the Mercator projection was provided by Superintendent
      Tittmann on March 10, 1910, when he named a chart board." See Shalowitz,
      Shore and Sea Boundaries, 2:301–2.

91   "There is no practical difference . . ." : U.S. Coast and Geodetic Survey, *An-
      nual Report of the Superintendent,* 141.

91   "Some of our charts . . ." : U.S. Coast and Geodetic Survey, *Annual Report of
      the Director,* 26.

91   . . . have yet to be converted: See the short discussion of projections in U.S.
      Department of Commerce, National Oceanic and Atmospheric Administra-
      tion, National Ocean Service, *Nautical Chart User's Manual,* 2-9 through
      2-11. Until NOAA was formed in 1970, charts of the Great Lakes were main-
      tained by the U.S. Lake Survey, part of the Army Corps of Engineers. As of
      summer 2002, a few of the Great Lakes charts were still on polyconic projec-
      tions.

91   . . . digital measurement technology . . . : Robert Wilson, Cartographic Re-
      search Division, National Ocean Survey, telephone conversation with au-
      thor, August 9, 2001.

91   In 1920 they developed . . . : U.S. Coast and Geodetic Survey, *Annual Report
      of the Director,* 35. For analysis of the projection's properties, see Deetz and
      Adams, *Elements of Map Projection,* 5th ed., 82–85.

92   Scale was not constant . . . : A scale bar that stood for one hundred miles along
      one of the standard parallels actually encompassed 100.6 miles at 39° N (mid-
      way between the secant lines) but only 99.0 miles at 39° N (along the north-
      ernmost border of the conterminous states), and 97.8 miles at 25° N (near the
      tip of Florida); see Stewart and Pierce, *Marine and Air Navigation,* 40.

92   "contact piloting" : For a discussion of different modes of air navigation, see
      Bryan, "Aeronautical Charts," 349.

94   "a conformal projection on which . . ." : International Civil Aviation Organi-
      zation, *Aeronautical Chart Manual,* 7-10-2.

94   A polar gnomonic might seem . . . : See, for example, Beresford, "Map Projec-
      tions."

95   . . . for tropical areas . . . : Guidelines called for adjusting the scale of the pro-
      jection so that the scale at 22.5° matched one of the commission's standard-
      ized round-number scales, such as 1:10,000,000 .

Chapter Seven: Soldiering On

General Sources

My key sources about military grids were Robison, "Military Grids"; and Skop,
"Evolution of Military Grids." For details about the transverse Mercator projection
and grid systems on U.S. Geological Survey maps, my principal sources were Sny-
der, *Flattening the Earth*; and Snyder, *Map Projections Used by the U.S. Geological
Survey*. Principal references for the Universal Transverse Mercator grid include
Colvocoresses, "Unified Plane Co-ordinate Reference System"; Hough, "Confor-
mal and World-Wide Military Grid System"; and O'Keefe, "Universal Transverse
Mercator Grid."

Notes

98   . . . without updating its map collars: According to John Snyder, map projec-
      tion guru at the Geological Survey in the 1980s, "The change of projections
      was further complicated by the fact that the projection identification in the
      map legend was not always changed with the projection. As a result, a map
      labeled as Polyconic that was produced during the late 1950s or early 1960s
      may be a Polyconic with updated map information, or it may actually be a
      Transverse Mercator or Lambert Conformal Conic." Snyder, "Differences
      Due to Projection," 200.

99   The mobile, rapid-fire . . . : For a description of the French 75, see Johnson,
      Artillery, 49–51; and U.S. Ordnance Department, *Handbook of Artillery,* 73–
      85. Licensed to French allies, the gun's hydraulic recoil system remained a
      military secret after the war.

99   The increased range . . . : For a concise assessment of the French 75's effect
      on military practice, see Dastrup, *Field Artillery,* 41–48.

100 I found several classic examples . . . : Flexner and Walker, *Military and Naval Maps and Grids,* quotations on 80–81.

100 . . . seldom align perfectly . . . : A simple explanation of this lack of alignment lies in the enforced parallelism of grid lines, which never meet, in contrast to the map's meridians, which converge, slowly but inevitably, toward the central meridian.

102 In the mid-1930s the U.S. Coast and Geodetic Survey . . . : Adams, "Flatland."

103 Figure 7.4: For the Lambert conformal conic projection the U.S. Geological Survey recommends standard parallels at 33° and 45° N for maps of the conterminous United States and at 20° and 60° N for maps of North America; see Snyder, *Map Projections Used by the U.S. Geological Survey,* 102.

104 . . . get a fix on true north . . . : O'Keefe, "Universal Transverse Mercator Grid," 22.

104 . . . their German foes, who had little appreciation . . . : See Skop, "Evolution of Military Grids," 16.

105 . . . local grids based on the Cassini projection . . . : See Chasseaud, "German Maps and Survey," esp. 126.

105 The infamous "Paris Gun" . . . : For discussion of the gun's operation and impact, see Miller, *Paris Gun.*

105 . . . based on an oblate globe . . . : An oblate globe, or spheroid, is described by rotating an ellipse about its shorter (minor) axis.

105 Historians trace . . . : Newton, *Principia,* 347–50, 821–26.

105 After nineteenth-century geodesists . . . : Bomford, *Geodesy,* 449–51; and Smith, *Introduction to Geodesy,* esp. 18–34.

106 . . . Gauss introduced the term *conformal*: Snyder, *Flattening the Earth,* 65.

106 . . . ellipsoidal tables for the polyconic projection . . . : Hunt and Schott, "Tables for Projecting Maps," 96–163.

106 . . . much is gained and little lost . . . : Aware of this advantage, Germany and Britain adopted the transverse Mercator projection for their military grids around 1920 and 1935, respectively.

106 "seriously impair . . ." : Skop, "Evolution of Military Grids," 16.

107 . . . declared the scale error "negligible" . . . : See Bowie and Adams, *Grid System for Progressive Maps,* quotation on 7. To avoid paper distortion, artillery officers requiring a graphic solution used grid coordinates to transfer positions from the map to a dimensionally stable medium with a printed or inscribed square grid; see O'Keefe, "Universal Transverse Mercator Grid," 22.

107 "unfortunate" . . . "azimuth errors . . ." : Skop, "Evolution of Military Grids," 17.

107 In 1947 the army rectified . . . : In developing the UTM system, the army relied heavily on advice from NATO allies as well as Oscar Sherman Adams (1874–1962), a senior Coast and Geodetic Survey mathematician. Although

he had helped Bowie with the ill-advised Polyconic Grid, Adams became a leading authority on map projection in the 1920s and supervised implementation of the State Plane Coordinate system in the 1930s. Roughly half of SPC zones employ the transverse Mercator projection, which the army decided to adopt in 1944, when Adams retired. Especially fond of conformal projections, Adams devised conformal transformations of the earth into a rectangle, a rhombus, a six-pointed star, and an ellipse. See Dracup, *Geodetic Surveys*; and Snyder, *Flattening the Earth,* 266.

107   . . . interpolar area between 84° N and 80° S . . . : The Universal Polar Stereographic (UPS) grid, based on a pair of pole-centered stereographic projections, covers regions north of 84° N and south of 80° S, essentially the north polar sea and Antarctica. For descriptions of UTM and SPC zone numbering and rectangular coordinates, see U.S. Department of the Army, *Grids and Grid References.*

107   "the evil effects . . ." : O'Keefe, "Universal Transverse Mercator Grid," 23. In addition, the six-degree zones conveniently coincided with sheets of the 1:1,000,000 International Map of the World.

108   "the property of orthomorphism . . ." : Hinks, *Map Projections,* 1st ed., v.

108   "the requirements of the artillery . . ." : Hinks, *Map Projections,* 2nd ed., viii.

CHAPTER EIGHT: ON TRACK

General Sources

Snyder's *Flattening the Earth* includes concise histories of the oblique Mercator and Space Oblique Mercator projections. A fuller description of the latter's mathematical properties appears in Snyder, *Space Oblique Mercator Projection: Mathematical Development.*

Notes

112   Perhaps the earliest use . . . : Snyder, *Flattening the Earth,* 95–96, 107, 161–62; and Eisele, "Charles S. Peirce," esp. 305–6. Snyder also notes the diverse mathematical approaches to the oblique Mercator projection by Max Rosenmund in 1903, Jean Leborde in 1926, J. H. Cole in 1943, and Martin Hotine in 1946. Hotine's version was used in designing the State Plane Coordinate grid for the Alaskan panhandle.

112   In March 1921 . . . : Smith, "From London to Australia by Aëroplane," map on 230–31. For the claim of priority, see Chamberlin, *Round Earth on Flat Paper,* 85–92.

113   Complementing Smith's . . . : Mitchell, "America in the Air."

113   By holding scale distortion . . . : For a description of the chart's properties, see U.S. Coast and Geodetic Survey, "New Aeronautical Route Chart."

116 . . . patented MapSat . . . : On February 2, 1982, the U.S. Patent Office awarded Colvocoresses patent no. 4,313,678 for "Automated Satellite Mapping System (MAPSAT)," filed on September 24, 1979, and assigned to "the United States of America, as represented by the Secretary of the Interior."

117 "era of automated mapping . . ." : Colvocoresses, "Space Oblique Mercator," 925.

117 . . . $22,000 on consultants' fees . . . : Ahmetaj, "John P. Snyder and Map Projections."

118 "provided the sought-after . . ." : Colvocoresses, foreword to Snyder, *Space Oblique Mercator Projection: Mathematical Development,* iv.

119 "I figured I was the last . . ." : Quotations from John Noble Wilford, "Map Hobbyist Sharpens Images of Earth from Space," *New York Times,* May 20, 1978.

120 "not . . . exactly conformal" : Snyder, "Space Oblique Mercator Projection," 586.

120 "would not have undertaken . . ." : Ibid., 596.

CHAPTER NINE: WALL MAPS AND WORLDVIEWS

General Sources

My examination of the prominence of the Mercator world map in the nineteenth and twentieth centuries relies heavily on Schulten, *Geographical Imagination*; and Snyder, *Flattening the Earth.* Key sources for inventors' insights on the Goode, Miller, and Robinson projections are Goode, "The Homolosine Projection"; Miller, "Notes on Cylindrical World Map Projections"; and Robinson, "New Map Projection."

Notes

121 But there it was . . . : A more recent sighting, in summer 2002, was a Mercator wall map, clearly visible from the sidewalk, in a storefront Christian Science Reading Room in downtown Ithaca, New York.

122 "laid down according . . ." : London *Daily Post,* April 6, 1731, quoted in Babinski, "Henry Popple's *Map,*" note 13.

123 . . . some atlas publishers preferred . . . : Snyder, *Flattening the Earth,* 96.

124 . . . a prolific author . . . : Morse's map was inserted opposite the title page.

124 Sidney's development . . . : After the engraver inscribed lines and type in soft, easily worked wax, electroplating converted the image to a rigid printing plate; see Woodward, *All-American Map,* esp. 16–23.

124 . . . Snyder tabulated . . . : Snyder, *Flattening the Earth,* 105.

124 A similar tabulation . . . : Ibid., 180–81.

125 . . . survived into the twenty-first century . . . : See, for example, the *National Geographic Family Reference Atlas,* 216–21.

125 "Mercator, Globular, and . . ." : Hinks, *Map Projections,* 1st ed., 69.

125 A half-century later . . . : See Wong, "World Map Projections," esp. 68–71. Wong's survey is based on atlases readily available in the university library, the Map Room of the New York Public Library, and the map library at the American Geographical Society.

125 "the great distortion . . ." : Hinks, *Map Projections,* 1st ed., 29.

125 "responsible for many . . ." : Deetz and Adams, *Elements of Map Projection,* 1st ed., 146.

126 "the Mercator world map . . . on a world map" : Raisz, *General Cartography,* 87.

126 "unfamiliar and unconventional" : Wong, "World Map Projections," 26.

126 "mental hazard . . ." : "Maps: Global War Teaches Global Geography," 57.

126 "the time has come . . ." : *New York Times,* "Airplanes and Maps."

126 Historian Susan . . . : Schulten, *Geographical Imagination,* 227.

126 "map industry [that] consciously . . ." : Ibid., 239.

126 "overexposure" : Ibid., 139.

127 . . . time-zone map in the otherwise progressive . . . : *Rand McNally Atlas of the World* (Chicago: Rand McNally, 2001), xiv.

130 . . . the most significant inroads . . . : Wong, "World Map Projections," 71–73.

132 "It is quite impossible . . ." : Goode, *Goode's School Atlas,* ix.

132 "(1) It presents the entire . . ." : Ibid., x.

133 "Only one Mercator projection . . ." : Ibid., x.

133 "In all previous atlases . . ." : Ibid., x.

134 "inducing lasting . . ." : Robinson, "New Map Projection," 148.

134 " . . . greatly enlarge Antarctica" : Ibid., 149.

134 At least that's the story . . . : Snyder, *Flattening the Earth,* 216. Use of the Robinson projection for world maps in editions of Rand McNally's *New International Atlas* published in 1972, 1987, and 1996 contradict Snyder's observation that the company dropped the Robinson map in its "popular" atlases.

135 . . . a thirteen-paragraph press release: National Geographic Society, "National Geographic's Millennial Gift."

136 "different and more realistic . . . ," "better still," "the trusty Van der Grinten," and "hope[d] that its main legacy . . ." : Garver, "New Perspective on the World," first quotation on 911, others on 913.

136 . . . more than 550 newspapers . . . : Robinson, "Flattening the Round Earth," cited in Snyder, *Flattening the Earth,* 216.

137 "presents a more realistic . . ." : *National Geographic Atlas of the World,* 6th ed., 13.

137 . . . in the February 1905 issue . . . : Prepared by the navy's Hydrographic Of-

fice, the map included transoceanic routes and submarine cables. As an accompanying note suggested, "The chart can be easily detached from the Magazine and hung on the wall for more convenient use." See "Chart of the World," *National Geographic* 16 (1905): 87.

137   "more than 75 percent . . ." and "avoids the congestion . . ." : *National Geographic Atlas of the World,* 7th ed., sheet 3.

138   Cropping was seldom . . . : In framing their Van der Grinten world map NGS mapmakers cut slightly more off the bottom, apparently to free up space for labeling the Arctic Ocean. See, for example, the physical world map in the fifth edition of the *National Geographic Atlas of the World* (1981); the projection extends slightly northward from 140° E, 80° N yet never reaches 140° E, 80° S.

138   "Eurocentric" : For critiques of Europe-centered maps, see Black, *Maps and History,* 199–203; and Harley, *New Nature of Maps,* 66, 157.

139   . . . ostensibly racist overtones . . . : For a concise critique of racism in historical atlases, see John Rennie Short, *Representing the Republic,* esp. 231–32.

140   "long used in British schools . . ." and "made Canada look . . ." : Jenkins, "Maps That Charted the Distortions of History."

140   I obtained my copy of Cram's *Map of the World* in October 2002 in Columbus, Ohio, at the annual meeting of the North American Cartographic Information Society. Identified as map no. CD-8 by the George F. Cram Co., Indianapolis, Ind., it lacks a publication date, a common practice of map publishers eager to prolong a product's shelf life.

140   "the simple fact that . . ." : Harley, *New Nature of Maps,* 66.

141   . . . Americans' sense of isolation . . . : This interpretation is mentioned in Schulten, *Geographical Imagination,* 228, 231.

141   . . . Japanese attack on Pearl Harbor . . . : This interpretation is reported in Tyner, "Interactions of Culture and Cartography," esp. 460–61.

141   . . . the John Birch Society speaker . . . : For a news story with a black-and-white (but nonetheless telling) photograph, see Bonfatti, "John Birch Speaker Looks to Congress."

141   Proponents of air power . . . : For an overview, see Henrikson, "Maps, Globes, and the Cold War"; and Henrikson, "Map as an 'Idea.'"

141   "out-perestroika[ed] perestroika" : Wattenberg, "Real Shape of the World Changes."

141   "its expansion . . ." and "if Germany . . ." : Mackinder, "Geographical Pivot of History," 436. For further discussion, see Boggs, "Cartohypnosis," esp. 470–72; Schulten, *Geographical Imagination,* 136–40; and Tyner, "Interactions of Culture and Cartography," 461.

143   . . . Miller and Van der Grinten felt . . . : Snyder, *Flattening the Earth,* 274.

143 "master image" : The concept's relevance to Mercator's map is described in Vujakovic, "Mapping the War Zone," esp. 190–91. Vujakovic credits MacEachren, *How Maps Work,* 455, who in turn attributes the master-image concept to Myers, "Every Picture Tells a Story."

Chapter Ten: Size Matters

General Sources

Key references are Crampton, "Alternative Cartographies"; Crampton, "Cartography's Defining Moment"; Monmonier, "Peters Projection Controversy"; Robinson, "Arno Peters and His New Cartography"; John P. Snyder, *Flattening the Earth*; Vujakovic, "Arno Peters' Cult"; and Vujakovic, "Extent of Adoption."

Notes

147 "advocate of equality in all things" : *Times* (London), "Arno Peters."

147 "when Britain was the strongest . . ." : Peters, *Peters Atlas of the World,* v. Too radical for the Peters atlas, the decimal graticule appears only on the back endpaper.

147 Although Peters discussed his world map . . . : Loxton, "Peters Phenomenon."

147 Reporters received copies . . . : According to Derek Maling, "about 350 reporters" attended. He does not cite a source for this estimate, which seems inflated—has a news conference ever attracted 350 reporters? In the next sentence Maling mentions a pro-Peters news story in the Manchester *Guardian,* but that story does not include a headcount. Perhaps pro-Peters publicists promulgated the estimate. Who knows? A number of other academic writers mention the *Guardian* story, which the newspaper attributes to the *Los Angeles Times.* See Maling, "Minor Modification," 509; and Morris, "Dr. Peters' Brave New World."

148 "a remarkable example . . ." : Maling, "Minor Modification," 509.

148 "The whole ten-year wonderwork . . ." : Maling, "Personal Projections," 600.

148 "the remarkable discovery . . ." : Maling, "Peters' Wunderwerk," 154.

149 "equivalent to shifting . . ." : Maling, "Personal Projections," 600.

149 . . . Peters "confessed" in 1980 . . . : See Crampton, "Alternative Cartographies," 70–71. Crampton quotes and cites an unpublished manuscript by Maling, intended as a final chapter in the second edition of Maling's map projection textbook but not included because of "the events in Europe in the early 1990s." Crampton, who presumably obtained a copy of Maling's manuscript from its author, found it insightful and regretted its puzzling withdrawal.

149 "geographical features . . ." and "for showing the comparative . . ." : Gall, "Use of Cylindrical Projections," 121.

149 "All I would ask . . ." : Ibid., 123.

150 "the best of all known . . ." : Quotation is a translation of the title of Behrmann's article, "Die beste bekannte flächentreue Projektion der ganzen Erde."

151 "as a 'general service' map . . ." : Quoted in Maling, "Some Notes about the Trystan Edwards Projection," 94. Presaging his critique of Peters, Maling argued that Edwards's 1953 map, unlike the projection the architect described, was not strictly equal area. Even so, John Snyder willingly accepted the cylindrical projection secant at 37° 24′ as an equal-area map; see Snyder, *Flattening the Earth,* 277.

152 Other named members . . . : See Craster, "Some Equal-area Projections"; and Balthasart, "L'emploi de la projection cylindrique équivalente."

152 . . . German historian's admittedly flawed 1973 map . . . : Crampton, "Alternative Cartographies," 74.

152 Although Peters was probably . . . : Terms quoted in this paragraph are from Peters, *Die Neue Kartographie/The New Cartography,* 105–18. The bilingual title reflects the juxtaposition of German and English text on a two-column page.

152 "detach a [small] section . . ." and "longitudinal distortion . . ." : Ibid., 117 and 113. For a table rating Goode's, Mercator's, and Peters's maps, among others, on the ten "attainable map qualities," see ibid., 114.

152 "all points which exist at an equal distance . . ." : Ibid., 109.

152 "does not deform . . ." : Ibid., 118.

153 "permits the construction . . ." : Ibid., 113.

153 "can cope with specialist . . ." : Ibid., 118.

153 . . . challenged Peters's claims . . . : See, for example, Board of the German Cartographical Society, "So-called Peters Projection"; and Wagner, "Das neue Kartenbild des Herrn Peters."

153 . . . Peters's map on the front covers . . . : See, for example, the paperback editions published in 1980 and 1983, respectively, by MIT Press.

153 "rather than the more familiar . . . ," " . . . innovative characteristics," and "represent[ing] an important step . . ." : Independent Commission on International Development Issues, *North–South,* 2.

153 . . . distributed over 60 million copies: "World Turned Upside Down."

154 " . . . statement in itself" : Vujakovic, "Extent of Adoption," 14.

154 "Philosophers, astronomers, historians . . ." : Peters, *Die Neue Kartographie/ The New Cartography,* 149.

155 Particularly offensive was . . . : A footnote at the end of the article attributes it to "*The Bulletin,* Press and Information Office of the Government of the Federal Republic of Germany, Bonn, Vol. 25, No. 17, Aug. 17, 1977, pp. 126–127."

156 "Map projections are fascinating . . .": Quotations of Robinson are from "American Cartographers."

157 "reads much more like . . .": This and other Snyder quotations in this paragraph, ibid.

157 "The Dr. Arno Peters' projection was acclaimed . . .": Ibid.

157 "Support for Professor Peters' map . . .": Kaiser, *New View of the World,* 9–10.

158 "Thus the mathematical or scientific superiority . . .": Ibid., 30–31.

158 "CORRECTION . . .": Ibid., insert pasted onto p. iii.

158 "Every teacher who really wants . . .": O'Driscoll, "Global Perspective."

158 " . . . predominantly white-dominated . . .": Sachs, "Cutting the Old World Down to Size," 16.

158 "no flat map will do . . .": Ibid., 17.

159 "Well, one needs to ask what . . .": Kaiser's February 3, 1984, radio interview with "All Things Considered" (National Public Radio) host David Malpus is quoted in Snyder, "Social Consciousness and World Maps," 192.

159 "wet, ragged, long, winter underwear . . .": Robinson, "Arno Peters and His New Cartography," 104.

159 "To force the spherical globe . . ." and "To a designer a rectangle . . .": Robinson, *Which Map Is Best?* 9. The other booklets, *Choosing a World Map: Attributes, Distortions, Classes, Aspects* and *Matching the Map Projection to the Need,* were published in 1988 and 1991, respectively.

159 "strongly urg[ing] book and map publishers . . .": All quotations in this paragraph are from the "Resolution Regarding the Use of Rectangular World Maps," included in American Cartographic Association, Committee on Map Projections, "Geographers and Cartographers Urge End"; full text of resolution on 223.

160 The implication was clear . . . : A letter from Arthur Robinson to John Snyder supports this interpretation. In sending the worked-over resolution back to Snyder, Robinson comments: "Enclosed is the next draft revision. Your turn. It is certainly long enough now. I think the primary objective is to bring the Mercator world map into disrepute, with a secondary aim of enveloping all rectangular world maps in the reaction, and I hope, some of that rubbing off on Gall-Peters." Robinson to Snyder, July, 8, 1988, John Snyder Collection, Geography and Map Division, U.S. Library of Congress, box 8, folder 5.

160 . . . antirectangular resolution won endorsements . . . : For reasons not readily apparent, the North American Cartographic Information Society, a rival organization, "declined to endorse it," and the Canadian Association of Geographers "did not respond." See Robinson, "Rectangular World Maps—No!" 103.

160 . . . mailing to three hundred news organizations . . . : Ibid.

160 The *Wall Street Journal* ran . . . : See *Wall Street Journal*, "Drawing the Line," June 8, 1989; and Booth, "Whereas the Earth Is Round."

160 . . . very little play in the press: A search of LexisNexis failed to turn up any other major newspapers or magazines that reported the resolution.

160 "the hysteria among leading cartographers . . ." : Harley, "Deconstructing the Map," 5.

160 "real issue," "agenda was the empowerment . . . ," and "whose power . . ." : Harley, "Can There Be a Cartographic Ethics?" 11.

160 "still closing ranks" : Ibid.

161 "The scientific Renaissance . . ." : Harley, "Deconstructing the Map," 6.

161 . . . found on all inhabited continents . . . : Saarinen, "Centering the Mental Maps."

161 . . . natural tendency to position oneself near the center: World maps centered on one's location are at least as old as 600 BC, when an unknown Babylonian cartographer inscribed a clay tablet with a world map centered on Babylon; see Millard, "Cartography in the Ancient Near East," esp. 111–12 and fig. 6.10 on 114.

161 "Cartography shares and reproduces . . ." : Pickles, "Texts, Hermaneutics and Propaganda Maps," 226.

162 "Enclosed is an *equal-area* pseudocylindrical-type projection . . ." : Snyder to Robinson, June 14, 1986, John Snyder Collection, Geography and Map Division, U.S. Library of Congress, box 8, folder 1.

162 "fairness to all peoples" : See, for example, ODT, Inc., "Characteristics of the Peters Map."

162 For an enlightening look at area cartograms, see Tobler, "Geographic Area and Map Projections."

162 I borrowed the armadillo graticule . . . : See Woytinsky and Woytinsky, *World Population and Production,* 42–43; and Raisz, Orthoapsidal World Maps." Also see Monmonier, "Originality Bites."

164 "Peters struck a chord . . ." : Black, *Maps and Politics,* 35.

164 "angry facts" : Brand, review of *The State of the World Atlas.*

164 "Since the world is virtually spherical . . ." : Kidron and Segal, *State of the World Atlas,* map 1.

165 "This Atlas represents . . ." : King and Vujakovic, "Peters Atlas," 245.

166 "interesting and varied" and "'no interpretation . . .'" : Vujakovic, "Arno Peters' Cult," 4–5.

166 "a distinct alternative . . ." : Hammond 2003 college catalog, 3.

166 . . . innovative "optimal conformal" projection . . . : A mathematical physicist better known for his contributions to fractal geometry and chaos theory, Feigenbaum devised Hammond's Optimal Conformal projection for the

Hammond Atlas of the World, published in 1993. Details are elusive, apparently because the projection is treated as a trade secret. For a concise description of Feigenbaum's rationale, see *Hammond Atlas of the World,* 10–11; Feigenbaum, "Using Nonlinear Dynamics"; and Snyder, *Flattening the Earth,* 246. John Snyder considered it "an encouraging computer-age advance in commercial projection choice."

168   "the largest selection . . .": See the Peters Map Seminar Pack page at the HR Press Web site, http://www.hrpress-diversity.com/peters.html.

168   . . . another diversity-awareness publisher . . .: See ODT, "New Map Resources."

168   "has shaken up cartography . . .": Kaiser and Wood, *Seeing through Maps,* 153 (inside back cover).

168   ODT's success . . .: According to an ODT press release, producer Aaron Sorkin, who got the idea for the show while surfing the Web, contacted the firm for examples and advice; see ODT, "President Carter's Nobel Prize."

168   Inquiries and sales leaped: See, for example, Abejo, "Sales Skyrocket"; and Johnson, "'West Wing' Piques Interest."

168   "The Mercator was designed . . .": NCC News Service, "NCC Friendship Press Peters Projection Map."

169   The projection's name . . .: ODT, "Press Room."

169   Although south-up maps . . . are not new . . .: For examples, see Black, *Maps and Politics,* 37–39; Ramphal, "World Turned Upside Down"; and "Upside Down, Inside Out." For a Web site with various examples, see Irving, "Upsidedown Map Page."

169   "When President Jimmy Carter receives . . .": ODT, "President Carter's Nobel Prize."

171   A good example can be found . . .: Knox and Marston, *Places and Regions in Global Context,* esp. 29–31, 121. For further discussion of polyhedral map projections, see Fuller, "Re-mapping Our World"; Leslie, "Energetic Geometries"; and White and others, "Comparing Area and Shape Distortion."

Chapter Eleven: Points of View

Notes

174   A case in point . . .: Harley, "Can There Be a Cartographic Ethics?" esp. 10–11.

174   "cartographic silences": For a development of this concept, see Harley, "Silences and Secrecy."

174   . . . his willingness to excuse . . .: See Harley, "Can There Be a Cartographic Ethics?" 10–11.

176   The Sunshine State's panhandle . . .: Bill Cleveland uses this example to

question judgments of area on maps. Another example is Oklahoma, which looks deceptively larger than Kansas. See Cleveland, *Elements of Graphing Data,* 282–84.

177 Critical theorists suspicious of government . . . : For a more detailed deconstruction of cartographic deconstructionists, see Black, *Maps and Politics,* 17–28.

177 "psychological isolation" . . . "a sort of 'Maginot Line' . . ." : Schulten, *Geographical Imagination,* 228.

177 "replac[ement of] the massive ocean buffers . . ." : Ibid., 231.

178 . . . recent biography . . . : Crane, *Mercator.*

178 "not linger[ing] on the social . . ." and other quotations in this paragraph are from Winchester, review of *Mercator.*

179 "The civilian flyer over a country . . ." : Stewart, "Use and Abuse of Map Projections," 601.

179 " . . . although a Mercator map of the world . . ." : Consolidated Vultee Aircraft Corporation, *Maps,* 16.

179 Computers that can fly the plane . . . : The principal limitation of GPS-based aviation is during take-off and landing; see Hofmann-Wellenhof, Lichtenegger, and Collins, *Global Positioning System,* 320–21.

179 . . . electronic aids for mariners . . . : A recent textbook on GPS-based marine navigation downplays rhumb lines in its detailed discussion of waypoint navigation. Nautical charts are useful for reading or plotting positions and avoiding hazards, but the Mercator projection's conformality is now much more important than its straight lines of constant bearing. See Monahan and Douglass, *GPS Instant Navigation,* esp. 148–49.

180 . . . John Birch Society propaganda . . . : For a newspaper photograph illustrating the group's use of the Mercator map at a public lecture, see Bonfatti, "John Birch Speaker Looks to Congress."

181 Figure 11.4: For a concise discussion of the commemorative issue, see Stage, "Stamps Mark Olympics, WWII."

181 "in the consideration of the various evils . . . ," "charts having correct areas . . . ," and "the Mercator projection, not only is a fixture . . ." : Deetz and Adams. *Elements of Map Projection,* 1st ed., 147.

182 "the standard map . . ." and "shape is much more . . ." : Luyten, "Those Misleading New Maps."

182 "All of us who know our geography . . ." : Ibid., 447–48. He identifies a few "uninformed amateurs" by name, along with their transgressions, in Luyten, "Air-Age Teaching or Misinforming?"

182 "moral preening" : Sowell, *Vision of the Anointed,* 147.

182 "deep thinkers . . ." and "Don't mess with . . ." : Sowell, "Drawing a Line."

183 "virtual globes" : A typical software application is Microsoft's Encarta World
     Atlas, marketed briefly in the late 1990s under the name "Virtual Globe."
     Users can turn the globe, view it from any angle, and zoom in or out. For re-
     views of a recent version, see DeBry and Richardson, "Encarta: The Refer-
     ence Tool of the Future?"; and "Best Atlases on CD-ROM." For a conceptual
     framework useful in selecting appropriate projections for printed maps, see
     Snyder, *Map Projections,* 33 – 35.

Bibliography

Abbe, Cleveland. "Comprehensive Maps and Models of the Globe for Special Meteorological Studies." *Monthly Weather Review* 35 (1907): 559–64.

Abejo, Jerry. "Sales Skyrocket for Map Shown on 'West Wing.'" *Gazette* (Montreal), March 6, 2001.

Adams, Oscar S. "Flatland: Not a Romance but a Necessary Expedient." *Journal of the Washington Academy of Sciences* 24 (1934): 201–16.

———. *The Study of Map Projections in General.* U.S. Coast and Geodetic Survey Special Publication 60. Washington, D.C.: Government Printing Office, 1919.

Admiralty (U.K.), Hydrographic Department. "Projections in Hydrographic Surveys: Adoption of Transverse Mercator." Hydrographic Department Professional Paper 15. London, 1953.

"Aeronautical Charts and the Oblique Mercator Projection." *Geographical Review* 37 (1947): 674–75.

Agnese, Battista. *Portolan Atlas of Nine Charts and a World Map* (ca. 1954). U.S. Library of Congress, "American Memory: Historical Collections for the National Digital Library." http://memory.loc.gov/.

Ahmetaj, Patty. "John P. Snyder and Map Projections." *ACSM* [American Congress on Surveying and Mapping] *Bulletin,* no. 167 (May/June 1997): 45–46.

Akerman, James R. "Atlas: Birth of a Title." In *The Mercator Atlas of Europe: Facsimile of the Maps of Gerardus Mercator Contained in the Atlas of Europe, circa 1570–1572,* edited by Marcel Watelet, 15–28. Pleasant Hill, Oreg.: Walking Tree Press, 1998.

"American Cartographers Vehemently Denounce German Historian's Projection."

ACSM [American Congress on Surveying and Mapping] *Bulletin,* no. 60 (February 1978): 27.

American Cartographic Association, Committee on Map Projections, "Geographers and Cartographers Urge End to Popular Use of Rectangular Maps," *American Cartographer* 16 (July 1989): 222–23.

Andrews, H. J. "Note on the Use of the Oblique Cylindrical Orthomorphic Projection." *Geographical Journal* 86 (1935): 446.

Averdunk, H., and J. Müller-Reinhard. "Gerard Mercator und die Geographen unter seinen Nachkommen." *Petermanns Geographische Mitteilungen,* suppl. no. 182 (1914).

Babinski, Mark. "Henry Popple's *Map of the British Empire in North America* (London, 1733)," ed. Matthew H. Edney. Osher Map Library and Smith Center for Cartographic Education, http://www.usm.maine.edu/maps/popple/.

———. *Henry Popple's 1733 Map of the British Empire in North America.* Garwood, N.J.: Krinder Peak Publishing, 1998.

Bagrow, Leo. *History of Cartography,* 2nd ed., revised and enlarged by R. A. Skelton. Chicago: Precedent Publishing, 1985. Originally published London: Watts, 1964; Cambridge, Mass.: Harvard University Press, 1966.

Bahn, Catherine I. "World Aeronautical Charts." *Bulletin, Special Libraries Association, Geography and Map Division,* no. 29 (1957): 15–18.

Balthasart, M. "L'emploi de la projection cylindrique équivalente dans l'enseignement de la géographie." *Société Belge d'Études Géographiques, Bulletin* 5 (1935): 269–72.

Barber, Peter M. "The British Isles." In *The Mercator Atlas of Europe: Facsimile of the Maps of Gerardus Mercator Contained in the Atlas of Europe, circa 1570–1572,* edited by Marcel Watelet, 43–77. Pleasant Hill, Oreg.: Walking Tree Press, 1998.

Bauer, L. A. "Halley's Earliest Equal Variation Chart." *Terrestrial Magnetism* 1 (1896): 28–31.

Behrmann, Walter. "Die beste bekannte flächentreue Projektion der ganzen Erde." *Petermanns Geographische Mitteilungen* 56, pt. 2, no. 3 (September 1910): 141–44.

Beresford, P. C. "Map Projections Used in Polar Regions." *Journal of Navigation* 6 (1953): 29–37.

"The Best Atlases on CD-ROM." *Booklist* 95 (1999): 916.

Bjerknes, V. "Sur les projections et les échelles a choisir pour les cartes géophysiques." *Geografiska Annaler* 2 (1920): 1–12.

Black, Jeremy. *Maps and History: Constructing Images of the Past.* New Haven, Conn.: Yale University Press, 1997.

———. *Maps and Politics.* London: Reaktion Press, 1997.

Blackmore, R. H. "Commercial Oversea Aviation Routes." *Proceedings of the United States Naval Institute* 66 (1940): 1427–38.

———. "Grid Navigation and Its Allied Problems." *Journal of Navigation* 1 (1948): 161–74.

Board of the German Cartographical Society. "The So-called Peters Projection." *Cartographic Journal* 22 (1985): 108–10.

Boggs, S. W. "Cartohypnosis." *Scientific Monthly* 64 (1947): 469–76.

Bomford, G. *Geodesy,* 3rd ed. Oxford: Clarendon Press, 1971.

Bonfatti, John F. "John Birch Speaker Looks to Congress." *Syracuse Post-Standard,* February 21, 1979.

Boorstin, Daniel J. *The Discoverers.* New York: Random House, 1983.

Booth, William. "Whereas the Earth Is Round." *Washington Post,* May 15, 1989.

Bowditch, Nathaniel. *Bowditch for Yachtsmen: Piloting.* New York: David McKay, 1976.

Bowie, William, and Oscar S. Adams. *Grid System for Progressive Maps in the United States.* U.S. Coast and Geodetic Survey Special Publication 59, revised. Washington, D.C.: Government Printing Office, 1919.

Bradford, Gerhsom. *The Whys and Wherefores of Navigation.* New York: D. Van Nostrand, 1919.

Brand, Steward. Review of *The State of the World Atlas,* by Michael Kidron and Ronald Segal. Global Business Network Book Club, January 1992. http://www.gbn.org/.

Brannon, Gary. "The Artistry and Science of Map-Making." *Geographical Magazine* 61 (September 1989): 38–40.

Brown, Lloyd. *The Story of Maps.* Boston: Little, Brown, 1949.

Bruce, A. P. C., and William Cogar. *An Encyclopedia of Naval History.* New York: Facts on File, 1998.

Bryan, G. S. "Aeronautical Charts." *Proceedings of the United States Naval Institute* 68 (1942): 349–56.

Burstyn, Harold L. "Maury, Matthew Fontaine." In *Dictionary of Scientific Biography,* edited by Charles Coulston Gillispie, 9:195–97. New York: Scribner, 1974.

Cajori, Florian. *The Chequered Career of Ferdinand Rudolph Hassler.* Boston: Christopher Publishing House, 1929; New York: Arno Press, 1980.

———. *A History of Mathematics,* 2nd ed. New York: Macmillan, 1938.

———. "On an Integration Ante-dating the Integral Calculus." *Bibliotheca Mathematica,* 3rd ser., 14 (1915): 312–19.

Calcoen, Roger, and others. *Le cartographie Gerard Mercator, 1512–1594.* Bruxelles: Crédit Communal, 1994.

Campbell, Tony. *Early Maps.* New York: Abbeville Press, 1981.

———. "Portolan Charts from the Late Thirteenth Century to 1500." In *The His-*

tory of Cartography, vol. 1, *Cartography in the Prehistoric, Ancient, and Me-dieval Europe and the Mediterranean,* edited by J. B. Harley and David Wood-ward, 371–463. Chicago: University of Chicago Press, 1987.

Carslaw, H. S. "The Story of Mercator's Map: A Chapter in the History of Mathe-matics." *Mathematical Gazette* 12 (January 1924): 1–7.

Carter Center. "Activities by Country" [map]. http://www.cartercenter.org/activities/activities.asp?submenu=activities.

Chamberlin, Wellman. *The Round Earth on Flat Paper: Map Projections Used by Cartographers.* Washington, D.C.: National Geographic Society, 1947.

Chapman, Sidney, and Julius Bartels. *Geomagnetism.* Oxford: Clarendon Press, 1940.

Chasseaud, Peter. "British Artillery and Trench Maps on the Western Front, 1914–1918." *Map Collector,* no. 51 (Summer 1990): 24–32.

———. "German Maps and Survey on the Western Front, 1914–1918." *Carto-graphic Journal* 38 (2001): 119–34.

———. "Mapping for D-Day: The Allied Landings in Normandy, 6 June 1944." *Cartographic Journal* 38 (2001): 177–89.

Chichester, Francis. "What Is the Ideal Air Map?" *Journal of Navigation* 1 (1948): 66–68.

Cleveland, William S. *Elements of Graphing Data.* Monterey, Calif.: Wadsworth, 1985.

"The Coast Survey." *Harper's New Monthly Magazine* 58 (1879): 506–21.

Cohen, Paul E., and Robert T. Augustyn. "A Newly Discovered Hondius Map." *The Magazine: Antiques* 155 (January 1999): 214–17.

Colvocoresses, Alden P. "Space Oblique Mercator." *Photogrammetric Engineering* 40 (1974): 921–26.

———. "A Unified Plane Co-ordinate Reference System." *World Cartography* 9 (1969): 9–65.

Consolidated Vultee Aircraft Corporation. *Maps, and How to Understand Them.* New York, 1943.

Cosgrove, Denis E. *Apollo's Eye: A Cartographic Genealogy of the Earth in the Western Imagination.* Baltimore: John Hopkins University Press, 2001.

Couper, Alastair, and others. *The Conway History of Seafaring in the Twentieth Century.* Washington, D.C.: Brassey's, 2000.

Crampton, Jeremy W. "Alternative Cartographies: New Frontiers of Human Is-sues." Ph.D. diss., Pennsylvania State University, 1994.

———. "Cartography's Defining Moment: The Peters Projection Controversy, 1974–1990." *Cartographica* 31 (Winter 1994): 16–32.

Crane, Nicholas. *Mercator: The Man Who Mapped the Planet.* New York: Henry Holt, 2002.

Craster, J. E. E. "Some Equal-area Projections of the Sphere." *Geographical Journal* 74 (1929): 471–74.

Daly, Charles P. "On the Early History of Cartography, or What We Know of the Maps and Map-Making, Before the Time of Mercator." *Journal of the American Geographical Society* 11 (1879): 1–40.

Dastrup, Boyd L. *The Field Artillery: History and Sourcebook.* Westport, Conn.: Greenwood Press, 1994.

Davies, Charles. *Elements of Surveying and Navigation; With a Description of the Instruments and the Necessary Tables.* New York: A. S. Barnes and Co., 1848.

Davis, John. *The Voyages and Works of John Davis: the Navigator,* edited by Albert Hastings Markham. London: Hakluyt Society, 1880.

DeBry, Steven, and John V. Richardson Jr. "Encarta: The Reference Tool of the Future?" *Library Quarterly* 71 (2001): 261–69.

Deetz, Charles H., and Oscar S. Adams. *Elements of Map Projection with Applications to Map and Chart Construction,* 1st ed. U.S. Coast and Geodetic Survey Special Publication 68. Washington, D.C.: Government Printing Office, 1921.

———. *Elements of Map Projection with Applications to Map and Chart Construction,* 5th ed. U.S. Coast and Geodetic Survey Special Publication 68. Washington, D.C.: Government Printing Office, 1945.

Dekker, Elly. *Globes at Greenwich: A Catalogue of the Globes and Armillary Spheres in the National Maritime Museum.* Oxford: Oxford University Press, 1999.

De Smet, Antoine. *Les sphères terrestre et céleste de Gérard Mercator 1541 et 1551.* Bruxelles: Editions Culture et Civilisation, 1968.

Dickson, A. F. "Ocean Routing Charts." *Journal of Navigation* 15 (1962): 339–41.

Dilke, O. A. W. *Greek and Roman Maps.* Ithaca, N.Y.: Cornell University Press, 1985.

Dracup, Joseph F. *Geodetic Surveys in the United States: The Beginning and the Next One Hundred Years, 1807–1940.* NOAA Web site, http://www.ngs .noaa.gov/PUBS_LIB/geodetic_surveying_1807.html.

Dunn, Carlos R. "The Weather and Circulation of July 1958: Heavy Precipitation Associated with a Trough in Central United States." *Monthly Weather Review* 86 (1958): 268–76.

Eisele, Carolyn. "Charles S. Peirce and the Problem of Map-Projection." *Proceedings of the American Philosophical Society* 107 (1963): 299–307.

Ellis, Melvin Y., ed. *Coastal Mapping Handbook.* Washington, D.C.: Government Printing Office, 1978.

Englisch, Brigitte. "Erhard Etzlaub's Projection and Methods of Mapping." *Imago Mundi* 48 (1996): 103–23.

Feeman, Timothy G. "Conformality, the Exponential Function, and World Map Projections." *College Mathematics Journal* 32 (2001): 334–42.

Feigenbaum, Mitchell J. "Using Nonlinear Dynamics to Make a New World Atlas." In *Towards the Harnessing of Chaos,* ed. Masaya Yamaguti, 1–9. Amsterdam: Elsevier, 1994.

Fillmore, Stanley, and R. W. Sandilands. *The Chartmakers: The History of Nautical Surveying in Canada.* Ottawa: Department of Fisheries and Oceans, Canadian Hydrographic Service: 1983.

Fite, Emerson D., and Archibald Freeman. *A Book of Old Maps Delineating American History from the Earliest Days to the Close of the Revolutionary War.* Cambridge, Mass.: Harvard University Press, 1926; New York: Dover, 1969.

Fleming, J. A., ed. *Terrestrial Magnetism and Electricity.* New York: McGraw-Hill, 1939.

Flexner, William Welch, and Gordon L. Walker. *Military and Naval Maps and Grids: Their Use and Construction.* New York: Dryden Press, 1942.

Freer, T. "A New Aeronautical Plotting Chart." *Journal of Navigation* 6 (1953): 358–61.

Freer, T. St. B., and K. J. Irwin. "Proposals for a New Air Navigation Chart." *Journal of Navigation* 1 (1948): 66–80.

French, Josephine, and others, eds. *Tooley's Dictionary of Mapmakers,* rev. ed. Riverside, Conn.: Early World Press, 2001.

Fuller, R. Buckminster. "Re-mapping Our World." *Today's Education* 63 (November–December 1974): 40–44, 107, 109–10.

Gall, James. "Use of Cylindrical Projections for Geographical, Astronomical, and Scientific Purposes." *Scottish Geographical Magazine* 1 (1885): 119–23.

Gannett, Henry. "The Mother Maps of the United States." *National Geographic* 4 (1892): 101–16.

Garver, John B., Jr. "New Perspective on the World." *National Geographic* 174 (1988): 910–13.

———. "Seventy-five Years of Cartography: A Love Affair with Maps." *National Geographic* 178 (November 1990): 130–34.

"Gerard Mercator's Map of the World (1569) in the Form of an Atlas in the Maritiem Museum 'Prins Hendrik' at Rotterdam." *Imago Mundi,* suppl. no. 2 (1961).

Gleick, James. *Chaos: Making a New Science.* New York: Viking Penguin, 1987.

Godlewska, Anne Marie Claire. *Geography Unbound: French Geographic Science from Cassini to Humboldt.* Chicago: University of Chicago Press, 1999.

———. "Jomard: The Geographic Imagination and the First Great Facsimile Atlases." In *Editing Early and Historical Atlases,* edited by Joan Winearls, 109–35. Toronto: University of Toronto Press, 1995.

Goode, J. Paul, ed. *Goode's School Atlas,* 1st ed. Chicago: Rand McNally, 1923.

———. "The Homolosine Projection: A New Device for Portraying the Earth's Surface Entire." *Annals of the Association of American Geographers* 15 (1925): 119–25.

Green, David R. "Journalistic Cartography: Good or Bad? A Debatable Point." *Cartographic Journal* 36 (1999): 141–53.

Gregg, W. R., and I. R. Tannehill. "International Standard Projections for Meteorological Charts." *Monthly Weather Review* 65 (1937): 411–15.

Gridgeman, N. T., and M. Zuker. "Mercator, the Antigudermannian, and a Fluke." *Canadian Cartographer* 15 (June 1978): 50–57.

Griggs, Armond L. "The Background and Development of Weather Charts." *Bulletin, Special Libraries Association, Geography and Map Division,* no. 21 (October 1955): 10–13.

Hall, Elial F. "Gerard Mercator: His Life and Works." *Journal of the American Geographical Society* 10 (1878): 163–96.

Halley, Edmund. "An Early Demonstration of the Analogy of the Logarithmick Tangents to the Meridian Line or Sum of the Secants." *Philosophical Transactions of the Royal Society of London* 19 (1696): 202–14.

Hammond Atlas of the World. Maplewood, N.J.: Hammond Incorporated, 1993.

Hammond Compact Peters World Atlas: The Earth in True Proportion. Union, N.J.: Hammond World Atlas Corporation, 2002.

Hammond World Atlas Corporation. Hammond 2003 college catalog. Union, N.J.: Hammond World Atlas Corporation, 2002.

Harley, J. B. "Can There Be a Cartographic Ethics?" *Cartographic Perspectives,* no. 10 (Summer 1991): 9–16.

———. "Deconstructing the Map." *Cartographica* 26 (Summer 1989): 1–20.

———. *The New Nature of Maps: Essays in the History of Cartography.* Ed. Paul Laxton. Baltimore: Johns Hopkins University Press, 2001.

———. "Silences and Secrecy: The Hidden Agenda of Cartography in Early Modern Europe." *Imago Mundi* 40 (1988): 57–76.

Harrison, Richard Edes. *Look at the World: The Fortune Atlas for World Strategy.* New York: Alfred A. Knopf, 1944.

Harvey, G. A. H. "International Civil Aviation Organization." *Journal of Navigation* 1 (1948): 76–78.

Haselden, Thomas. *The Description and Use of That Most Excellent Invention Commonly Call'd Mercator's Chart.* London, 1722. Reprint, The Eighteenth Century [microfilm collection], 1166:3, Woodbridge, Conn.: Research Publications, 1985.

———. *A Reply to Mr. Wilson's Answer to My Letter to Dr. Halley.* London, 1722.

Reprint, The Eighteenth Century [microfilm collection], 1166:4, Woodbridge, Conn.: Research Publications, 1985.

Hassler, Ferdinand Rudolph. "Papers on Various Subjects Connected with the Survey of the Coast of the United States." *Transactions of the American Philosophical Society,* n.s., 2 (1825): 232–420.

Hayden, Everett. "The Pilot Chart of the North Atlantic Ocean." *Journal of the Franklin Institute* 125 (1888): 265–78, 447–62.

Heawood, Edward. *A History of Geographical Discovery in the Seventeenth and Eighteenth Centuries.* Cambridge: Cambridge University Press, 1912.

———. "Hondius and His Newly-Found Map of 1608." *Geographical Journal* 45 (1919): 178–84.

———. "Lost Mercator Maps." *Geographic Journal* 62 (1923): 138–40.

Hellemans, Alexander, and Bryan Bunch. *The Timetables of Science: A Chronology of the Most Important People and Events in the History of Science.* New York: Simon and Schuster, 1988.

Henrikson, Alan K. "The Map as an 'Idea': The Role of Cartographic Imagery during the Second World War." *American Cartographer* 2 (1975): 19–53.

———. "Maps, Globes, and the Cold War." *Special Libraries* 65 (1974): 445–54.

Herrick, Samuel. "Grid Navigation." *Geographical Review* 34 (1944): 436–56.

Heyer, Alfons. "Drei Mercator-Karten in der Breslauer Stadt-Bibliothek." *Zeitschrift für Wissenschaftliche Geographie* 7 (1889): 379–89, 474–87, 507–28.

Hilgard, J. E. "Table for Projecting Maps of Large Extent." Appendix 58 in *Report of the Superintendent of the Coast Survey, Showing the Progress of the Survey during the Year 1856,* 296–307. Washington, D.C.: A. O. P. Nicholson, Printer, 1856.

———. "Tables for Projecting Maps of Large Extent." Appendix 33 in *Report of the Superintendent of the Coast Survey, Showing the Progress of the Survey during the Year 1859,* 328–58. Washington, D.C.: Thomas H. Ford, Printer, 1860.

Hind, Arthur Mayger. *Engraving in England in the Sixteenth and Seventeenth Centuries: A Descriptive Catalogue with Introductions.* Cambridge: University Press, 1952.

Hinks, Arthur R. *Map Projections,* 1st ed. Cambridge: University Press, 1912.

———. *Map Projections,* 2nd ed. Cambridge: University Press, 1921.

Hobbs, Richard R. *Marine Navigation 1: Piloting.* Annapolis, Md.: Naval Institute Press, 1974.

———. *Marine Navigation 2: Celestial and Electronic.* Annapolis, Md.: Naval Institute Press, 1974.

Hodson, Yolanda. "MacLeod, MI4, and the Directorate of Military Survey, 1919–1943." *Cartographic Journal* 38 (2001): 155–75.

Hoff, Bert van 't. Introduction to "Gerard Mercator's Map of the World (1569) in the Form of an Atlas in the Maritiem Museum 'Prins Hendrik' at Rotterdam." *Imago Mundi,* suppl. no. 2 (1961): 1–69.

Hofmann-Wellenhof, B., H. Lichtenegger, and J. Collins. *Global Positioning System Theory and Practice,* 5th ed. New York: Springer, 2001.

Hooker, Brian. "New Light on Jodocus Hondius' Great World Mercator Map of 1598." *Geographical Journal* 159 (1993): 45–50.

Hough, Floyd W. "A Conformal and World-Wide Military Grid System." *Transactions of the American Society of Civil Engineers* 121 (1956): 633–41.

HR Press. Peters Map Seminar Pack. http://www.hrpress-diversity.com/ peters.html.

Hunt, E. B., and Charles A. Schott. "Tables for Projecting Maps, with Notes on Map Projections." Appendix 39 in *Report of the Superintendent of the Coast Survey, Showing the Progress of the Survey during the Year 1853,* 96–163. Washington, D.C.: Robert Armstrong, Public Printer, 1854.

Independent Commission on International Development Issues. *North–South: A Programme for Survival.* Cambridge, Mass.: MIT Press, 1980.

International Civil Aviation Organization. *Aeronautical Chart Manual,* 2nd ed. Montreal: International Civil Aviation Organization, 1987.

Irving, Francis. "The Upsidedown Map Page." http://www.flourish.org/ upsidedownmap/.

Jenkins, Simon. "Maps That Charted the Distortions of History." *Times* (London), August 3, 2001.

Johnson, Curt. *Artillery.* London: Octopus Books, 1975.

Johnson, Marylin. "'West Wing' Piques Interest in Unusual Map." *Atlanta Journal and Constitution,* March 8, 2001.

Johnson, William E. "The World Geographic Reference System." *Air University Quarterly Review* 4 (Summer 1951): 49–57.

Kaiser, Ward L. *A New View of the World: A Handbook to the World Map, Peters Projection.* New York: Friendship Press, 1987.

Kaiser, Ward L., and Denis Wood. *Seeing through Maps: The Power of Images to Shape Our World View.* Amherst, Mass.: ODT Incorporated, 2001.

Karrow, Robert W., Jr. Commentary accompanying Gerardus Mercator, *Atlas sive Cosmographica Meditationes de Fabrica Mundi et Fabricati Figura.* Duisburg, 1595; Reprint, Oakland, Calif.: Octavo, 2000. CD-ROM.

———. *Mapmakers of the Sixteenth Century and Their Maps.* Chicago: Speculum Orbis Press for the Newberry Library, 1993.

Kelley. J. E., Jr. "The Oldest Portolan Chart in the New World." *Terrae Incognitae* 9 (1977): 23–48.

Kemp, Peter, ed. *The Oxford Companion to Ships and the Sea.* Oxford: Oxford University Press, 1976.

Keuning, Johannes. "The History of Geographical Map Projections until 1600." *Imago Mundi* 12 (1955): 1–24.

Kidron, Michael, and Ronald Segal. *State of the World Atlas.* New York: Simon and Schuster, 1981.

King, Russell, and Peter Vujakovic. "Peters Atlas: A New Era of Cartography or Publisher's Con-Trick?" *Geography* 74 (1989): 245–51.

Kirmse, Rolf. "Die grosse Flandernkarte Gerhard Mercators (1540)—ein Politicum?" *Duisburger Forschungen* 1 (1957): 1–44.

Kish, George. "Mercator, Gerardus (or Gerhard Kremer)." In *Dictionary of Scientific Biography,* edited by Charles Coulston Gillispie, 9:309–10. New York: Scribner, 1974.

Knox, Paul, and Sallie Marston. *Places and Regions in Global Context: Human Geography.* Upper Saddle River, N.J.: Prentice-Hall, 1998.

Knox, R. W. "The Mercator Projection." *International Hydrographic Bulletin,* November 1969, 370–72.

Koeman, C. "Gerhard Mercator's Map of the World (1569), in the Form of an Atlas in the Maritiem Museum 'Prins Hendrik' at Rotterdam [review]." *Imago Mundi* 17 (1963): 115–16.

Kretschmer, Ingrid. "Kartenprojektion." In *Lexikon zur Geschichte der Kartographie,* 1:376–85. Vienna: Franz Deuticke, 1986.

———. "Mercators Bedeutung in der Projektionslehre (Mercatorprojektion)." In *Mercator und Wandlungen der Wissenschaften im 16. und 17. Jahrhundert,* edited by Manfred Büttner and René Dirven, 151–74. Bochum: Brockmeyer, 1993.

Krücken, Wilhelm. "Ad maiorem Gerardi Mercatoris gloriam." http://www.wilhelmkruecken.de/.

———. "Wissenschaftsgeschichtliche und -theoretische Überlegungen zur Entstehung der Mercator-Weltkarte 1569 AD USUM NAVIGANTIUM." *Duisburger Forschungen* 41 (1994): 1–92.

Lambert, J. H. *Notes and Comments on the Composition of Terrestrial and Celestial Maps (1772),* translated and introduced by Waldo R. Tobler. Michigan Geographical Publication 8. Ann Arbor: Department of Geography, University of Michigan, 1972.

Lanman, Jonathan T. *On the Origin of Portolan Charts.* Chicago: Newberry Library, 1987.

Laskowski, Piotr H. "The Traditional and Modern Look at Tissot's Indicatrix." In *The Accuracy of Spatial Databases,* edited by Michael Goodchild and Sucharita Gopal, 155–74. London: Taylor and Francis, 1989.

Lawrence, David M. *Upheaval from the Abyss: Ocean Floor Mapping and the Earth Science Revolution.* New Brunswick: Rutgers University Press, 2002.

Lecky, Squire Thornton Stratford. *Wrinkles in Practical Navigation, Together with a Brief Memoir of the Author by His Son.* New York: Van Nostrand, 1956.

Lee, L. P. "Nomenclature and Classification of Map Projections." *Empire Survey Review* 51 (1944): 190–200.

LeGear, Clara Egli. "Mercator's Atlas of 1595." *U.S. Library of Congress, Quarterly Journal of Current Acquisitions* 7 (May 1950): 9–13.

———. "Sixteenth-Century Maps Presented by Lessing J. Rosenwald." [U.S. Library of Congress] *Quarterly Journal of Current Acquisitions* 6 (May 1949): 18–22.

Leslie, Thomas W. "Energetic Geometries: The Dymaxion Map and the Skin/Structure Fusion of Buckminster Fuller's Geodesics." *ARQ: Architectural Research Quarterly* 5 (2001): 161–70.

Lewis, Charles Lee. *Matthew Fontaine Maury, the Pathfinder of the Seas.* Annapolis, Md.: United States Naval Institute, 1927.

Lohne, J. A. "Harriot (or Hariot), Thomas." In *Dictionary of Scientific Biography,* edited by Charles Coulston Gillispie, 6:124–29. New York: Scribner, 1972.

———. "Lambert, Johann Heinrich." In *Dictionary of Scientific Biography,* edited by Charles Coulston Gillispie, 7:595–600. New York: Scribner, 1973.

Loomer, Scott A. "A Cartometric Analysis of Portolan Charts: A Search Methodology." Ph.D. diss., University of Wisconsin–Madison, 1987. Abstract in *Dissertation Abstracts International* 48 (1987): 2136A.

———. "Mathematical Analysis of Medieval Sea Charts." *Technical Papers, 1986 ACSM-ASPRS Annual Convention,* 1:123–32.

López de Azcona, J. M. "Nuñes Salaciense, Pedro." In *Dictionary of Scientific Biography,* edited by Charles Coulston Gillispie, 10:160–62. New York: Scribner, 1974.

Loxton, John. "The Peters Phenomenon." *Cartographic Journal* 22 (1985): 106–8.

Luyten, Willem J. "Air-Age Teaching or Misinforming?" *Science* 97 (1943): 201–2.

———. "Those Misleading New Maps." *Harper's Magazine* 187 (October 1943): 447–49.

MacEachren, Alan M. *How Maps Work: Representation, Visualization, and Design.* New York: Guilford Press, 1995.

Mackinder, H. J. "The Geographical Pivot of History." *Geographical Journal* 23 (1904): 421–37.

Maling, Derek. "A Minor Modification to the Cylindrical Equal-Area Projection," *Geographical Journal* 140 (1974): 509–10.

———. "Personal Projections." *Geographical Magazine* 46 (August 1974): 599–600.

———. "Peters' Wunderwerk." *Kartographische Nachrichten* 24 (1974): 153–56.

———. "Some Notes about the Trystan Edwards Projection." *Cartographic Journal* 3 (1966): 94–97.

Maloney, Elbert S. *Dutton's Navigation and Piloting,* 13th ed. Annapolis, Md.: Naval Institute Press, 1978.

Manning, Thomas G. *U.S. Coast Survey vs. Naval Hydrographic Office: A 19th-Century Rivalry in Science and Politics.* Tuscaloosa: University of Alabama Press, 1988.

Maor, Eli. *Trigonometric Delights.* Princeton, N.J.: Princeton University Press, 1998.

"Maps: Global War Teaches Global Geography." *Life,* August 3, 1942, 57–64.

Marcano, Tony. "J. P. Snyder, 71; Used Satellites to Map Earth." *New York Times,* May 2, 1997.

Marschner, F. J. "Structural Properties of Medium- and Small-Scale Maps." *Annals of the Association of American Geographers* 34 (1944): 1–46.

Matthew, Donald. *Atlas of Medieval Europe.* New York: Facts on File, 1983.

Maurer, H. "Johann Heinrich Lambert." *International Hydrographic Review* 8 (May 1931): 70–82.

Maury, Matthew Fontaine. *The Physical Geography of the Sea and Its Meteorology.* New York: Harper and Brothers, 1856. Reprint, edited by John Leighly, Cambridge, Mass.: Harvard University Press, Belknap Press, 1963.

May, Kenneth O. "Gauss, Carl Friedrich." In *Dictionary of Scientific Biography,* edited by Charles Coulston Gillispie, 5:298–315. New York: Scribner, 1972.

McCaw, G. T. "The Transverse Mercator Projection: A Critical Examination." *Empire Survey Review* 5 (1940): 285–96.

McKinney, William M. "The Wright Projection." *Journal of Geography* 68 (1969): 472.

McMaster, Robert, and Susanna McMaster. "A History of Twentieth-Century American Academic Cartography." *Cartography and Geographic Information Systems* 29 (2002): 305–21.

Melluish, R. K. *An Introduction to the Mathematics of Map Projections.* Cambridge: University Press, 1931.

Mercator, Gerardus. *Atlas sive Cosmographicæ Meditationes de Fabrica Mundi et Fabricati Figura.* Duisburg, 1595; Reprint, Oakland, Calif.: Octavo, 2000. CD-ROM.

"Mercator's World Map of 1569." *Geographical Journal* 81 (1933): 479.

Meurer, Peter H. *Fontes cartographici Orteliani: das "Theatrum orbis terrarum" von Abraham Ortelius und seine Kartenquellen.* Weinheim: VCH, 1991.

Millard, A. R. "Cartography in the Ancient Near East." In *The History of Cartography,* vol. 1, *Cartography in the Prehistoric, Ancient, and Medieval Europe and the Mediterranean,* ed. J. B. Harley and David Woodward, 107–16. Chicago: University of Chicago Press, 1987.

Miller, Henry W. *The Paris Gun: The Bombardment of Paris by the German Long Range Guns and the Great German Offensives of 1918.* New York: Jonathan Cape and Harrison Smith, 1930.

Miller, O. M. "An Experimental Air Navigation Map." *Geographical Review* 23 (1933): 48–60.

———. "Notes on Cylindrical World Map Projections." *Geographical Review* 32 (1942): 424–30.

Mitchell, William. "America in the Air: The Future of Airplane and Airship, Economically and as Factors in National Defense." *National Geographic* 39 (1921): 339–52.

Mixter, George W. *Primer of Navigation,* 2nd ed. New York: D. Van Nostrand, 1943.

Mollat, Michel. *Sea Charts of the Early Explorers: 13th to 17th Century.* New York: Thames and Hudson, 1984.

Monahan, Kevin, and Don Douglass. *GPS Instant Navigation: From Basic Techniques to Electronic Charting,* 2nd ed. Anacortes, Wash.: FineEdge.com, 2000.

Monmonier, Mark. *Air Apparent: How Meteorologists Learned to Map, Predict, and Dramatize Weather.* Chicago: University of Chicago Press, 1999.

———. *How to Lie with Maps.* Chicago: University of Chicago Press, 1991.

———. "The Lives They Lived: John P. Snyder." *New York Times Magazine,* January 4, 1998, 33.

———. "Originality Bites: Copyright Registration and Map History." *Mercator's World* 6 (September/October 2001): 50–52.

———. "The Peters Projection Controversy." In *Drawing the Line: Tales of Maps and Cartocontroversy,* 9–44. New York: Henry Holt, 1995.

Moore, John Hamilton. *Seaman's Complete Daily Assistant, and New Mariner's Compass, Being an Easy Method of Keeping a Journal at Sea.* London: B. Law, 1796.

Morris, Joe Alex. "Dr. Peters' Brave New World." *Guardian* (Manchester), June 5, 1973, 15.

Morse, Jedidiah. *A Compendium and Complete System of Modern Geography, or a View of the Present State of the World.* Boston: Thomas and Andrews, 1814.

Mottelay, Paul F. *Bibliographical History of Electricity and Magnetism, Chronologically Arranged.* London: C. Griffin and Co., 1922.

Myers, Greg. "Every Picture Tells a Story: Illustrations in E. O. Wilson's *Sociobiology.*" In *Representation in Scientific Practice,* edited by Michael Lynch and Steve Woolgar, 231–65. Cambridge, Mass.: MIT Press, 1990.

National Geographic Atlas of the World, 6th ed. Washington, D.C.: National Geographic Society, 1990.

National Geographic Atlas of the World, 7th ed. Washington, D.C.: National Geographic Society, 1999.

National Geographic Family Reference Atlas, 2nd ed. Washington, D.C.: National Geographic Society, 2002.

National Geographic Society. "National Geographic's Millennial Gift to America: Every School to Receive an Updated World Map." Press release, September 8, 1998. http://www.nationalgeographic.com/events/releases/pr980908.html.

NCC (National Council of Churches) News Service. "NCC Friendship Press Peters Projection Map to be Featured on NBC-TV's 'West Wing,'" February 13, 2001. http://www.ncccusa.org/news/01news13.html.

Nebenzahl, Kenneth. *Atlas of Columbus and the Great Discoveries.* Chicago: Rand McNally, 1990.

———. *Maps from the Age of Discovery: Columbus to Mercator.* London: Times Books, 1990.

Newton, Isaac. *The Mathematical Papers of Isaac Newton.* Vol. 1, *1664–1666,* edited by D. T. Whiteside. Cambridge: Cambridge University Press, 1967.

———. *The Principia: Mathematical Principles of Natural Philosophy.* Translated by I. Bernard Cohen and Anne Whitman. Berkeley: University of California Press, 1999.

"A New View of the World," *Christianity Today* 28 (February 17, 1984): 39–40.

New York Times. Editorial, "Airplanes and Maps," February 21, 1943, sec. 4.

Nordenskiöld, Adolf Erik. *Facsimile Atlas to the Early History of Cartography with Reproductions of the Most Important Maps Printed in the XV and XVI Centuries,* translated by Johan Adolf Ekelöf and Clements R. Markham. Stockholm, 1889. Reprint, New York: Dover Publications, 1973.

O'Driscoll, Patrick. "The Global Perspective: Third World Gains." *USA Today,* Janaury 23, 1984, 9A.

ODT, Inc. "Characteristics of the Peters Map." http://www.petersmap.com/page9.html.

———. "New Map Resources." ODT Web site, http://www.odt.org/.

———. "President Carter's Nobel Prize." http://www.odt.org/cartermap.htm.

O'Keefe, John A. "The Universal Transverse Mercator Grid and Projection." *Professional Geographer* 4 (September 1952): 19–24.

Osley, A. S. *Mercator. A Monograph on the Lettering of Maps, Etc. in the 16th Century Netherlands with a Facsimile and Translation of His Treatise on the Italic Hand and a Translation of Ghim's "Vita Mercatoris."* New York: Watson-Guptill, 1969.

Paradis, Michael G. *The Mercator Projection and Its Variations.* Informal Report 67-37. Washington, D.C.: Naval Oceanographic Office, 1967.

Parsons, E. J. S., and W. F. Morris. "Edward Wright and His Work." *Imago Mundi* 3 (1938): 61–71.

Peck, Don. "The Gun Trade." *Atlantic Monthly* 290 (December 2002): 46–47.

Pepper, Jon V. "Harriot's Calculation of the Meridional Parts as Logarithmic Tangents." *Archive for History of Exact Sciences* 4 (1968): 359–413.

Peters, Arno. *Die Neue Kartographie/The New Cartography.* Klagenfurt, Austria: Universitätsverlag Carinthia; New York: Friendship Press, 1983.

———. *Peters Atlas of the World,* 1st U.S. ed. New York: Harper and Row, 1990.

Peters, Arno, and Anneliese Peters. *Synchronoptische Weltgeschichte.* Frankfurt am Main: Universum-Verlag, 1952.

"Peters Projection—to Each Country Its Due on the World Map." *ACSM* [American Congress on Surveying and Mapping] *Bulletin,* no. 59 (November 1977): 13–15.

Pickles, John. "Texts, Hermaneutics and Propaganda Maps." In *Writing Worlds: Discourse, Text and Metaphor in the Representation of Landscape,* ed. Trevor J. Barnes and James S. Duncan, 193–230. London: Routledge, 1992.

Pillsbury, John E. "Charts and Chart Making." *Proceedings of the United States Naval Institute* 10 (1884): 187–202.

Pinson, Stephen C. "Repressed Mimesis: Jomard and the *Monuments de la Géographie.*" *Portolan* [Washington Map Society], no. 38 (Spring 1997): 5–17.

Quinn, D. B., ed. *The Hakluyt Handbook.* Vol. 2. London: Hakluyt Society, 1974.

Raisz, Erwin. *General Cartography.* New York: McGraw-Hill, 1938.

———. "Orthoapsidal World Maps." *Geographical Review* 33 (1943): 132–34.

Ramphal, Shridath S. "A World Turned Upside Down." *Geography* 70 (1985): 193–205.

Reed, Ronald. *The Nature and Making of Parchment.* Leeds, U.K.: Elmete Press, 1975.

Reingold, Nathan. "Hassler, Ferdinand Rudolph." In *Dictionary of Scientific Biography,* edited by Charles Coulston Gillispie, 6:165–66. New York: Scribner, 1972.

Reynolds, Nigel. "First Maps of Britain Chart a Catholic Plot; British Library Buys Atlas after 20 Years." *Daily Telegraph,* May 29, 1997.

Rickey, V. Frederick, and Philip M. Tuchinsky. "An Application of Geography to Mathematics: History of the Integral of the Secant." *Mathematics Magazine* 53 (May 1980): 162–66.

Rigg, J. B. "Maps and the Meteorologist." *Weather* 12 (May 1957): 154–57.

Ristow, Walter W. *Aviation Cartography: A Historico-Bibliographic Study of Aeronautical Charts,* 2nd ed. Washington, D.C.: Library of Congress, 1960.

Robinson, A. H. W. "The Evolution of the English Nautical Chart." *Journal of Navigation* 5 (1952): 362–74.

Robinson, Arthur H. "Arno Peters and His New Cartography." *American Cartographer* 12 (1985): 103–11.

———. *Choosing a World Map: Attributes, Distortions, Classes, Aspects.* Falls Church, Va.: American Congress on Surveying and Mapping, American Cartographic Association, Committee on Map Projections, 1988.

———. "A New Map Projection: Its Development and Characteristics." *International Yearbook of Cartography* 14 (1974): 145–55.

———. "Rectangular World Maps—No!" *Professional Geographer* 42 (1990): 101–4.

———. *Which Map Is Best? Projections for World Maps.* Falls Church, Va.: American Congress on Surveying and Mapping, American Cartographic Association, Committee on Map Projections, 1986.

Robinson, Arthur H., and John P. Snyder, eds. *Matching the Map Projection to the Need.* Bethesda, Md.: American Congress on Surveying and Mapping, American Cartographic Association, Committee on Map Projections, 1991.

Robison, John B. "Military Grids: Theory, History and Utilization." U.S. Army Map Service, Far East, Papers on Cartography, 1958–59. Mimeograph available in the NOAA Central Library.

Ross, Raymond L. "The United States Sectional Airway Maps." *Military Engineer* 24 (1932): 273–76.

Rosser, W. H. "Mercator's Chart: Its History and Abuse." *Nautical Magazine and Naval Chronicle* 52 (1883): 105–11, 288–91; 53 (1884): 278–86, 448–51.

Saarinen, Thomas F. "Centering the Mental Maps of the World." *National Geographic Research* 4 (1988): 112–27.

Sachs, Dana. "Cutting the Old World Down to Size." *Mother Jones* 11 (December 1986): 16–17.

Schilder, Günter. "Willen Jansz. Blaeu's Wall Map of the World, on Mercator's Projection, 1606–07, and Its Influence." *Imago Mundi* 31 (1979): 36–54.

Schnelbögl, Fritz. "Life and Work of the Nuremberg Cartographer Erhard Etzlaub," *Imago Mundi* 20 (1966): 11–26.

Schnell, George A., and Mark Monmonier. *The Study of Population: Elements, Patterns, Processes.* Columbus, Ohio: Charles E. Merrill, 1983.

Schott, Charles A. "A Comparison of the Relative Value of the Polyconic Projection Used on [*sic*] the Coast and Geodetic Survey, with Some Other Projections." Appendix 10 in *Report of the Superintendent of the U.S. Coast and Geodetic Survey Showing the Progress of the Work during the Fiscal Year Ending with June, 1880,* 287–96. Washington, D.C.: Government Printing Office, 1882.

Schulten, Susan. *The Geographical Imagination in America, 1880–1950.* Chicago: University of Chicago Press, 2001.

———. "Richard Edes Harrison and the Challenge to American Cartography." *Imago Mundi* 50 (1998): 174–88.

Scott, Peter, and John Goss. "Important Mercator 'Discovery' under the Hammer." *Map Collector,* no. 6 (March 1979): 27–35.

Scriba, Christoph J. "Lambert, Johann Heinrich." In *Dictionary of Scientific Biography,* edited by Charles Coulston Gillispie, 7:595–600. New York: Scribner, 1973.

Shalowitz, Aaron L. "The Chart That Made Navigation History." *Journal of the Washington Academy of Sciences* 59 (October–December 1969): 180–86.

———. *Shore and Sea Boundaries, with Special Reference to the Interpretation and Use of Coast and Geodetic Survey Data,* 2 vols. Washington, D.C.: Government Printing Office, 1962–64.

Sheppard, H. L. "A Note on Chart Distortions." *Journal of Navigation* 6 (1953): 159–60.

Shirley, Rodney W. *The Mapping of the World: Early Printed World Maps, 1472–1700.* London: Holland Press, 1983.

Short, John Rennie. *Representing the Republic: Mapping the United States 1600–1900.* London: Reaktion Books, 2001.

Shute, William, and others. *An Introduction to Navigation and Nautical Astronomy.* New York: MacMillan, 1944.

Skop, Jacob. "The Evolution of Military Grids." *Military Engineer* 43 (1951): 15–18.

Smith, James R. *Introduction to Geodesy: The History and Concepts of Modern Geodesy.* New York: John Wiley and Sons, 1997.

Smith, Paul A. "Aeronautical Chart Production." *Military Engineer* 35 (1943): 357–61.

Smith, Ross. "From London to Australia by Aëroplane: A Personal Narrative of the First Aërial Voyage Half Around the World." *National Geographic* 39 (1921): 229–339.

Snyder, John P. "Differences Due to Projection for the Same USGS Quadrangle." *Surveying and Mapping* 47 (1987): 199–206.

———. *Flattening the Earth: Two Thousand Years of Map Projections.* Chicago: University of Chicago Press, 1993.

———. *Map Projections—A Working Manual.* Geological Survey Professional Paper 1395. Washington, D.C.: Government Printing Office, 1987.

———. "Map Projections for Satellite Tracking." *Photogrammetric Engineering and Remote Sensing* 47 (1981): 205–13.

———. *Map Projections Used by the U.S. Geological Survey.* U.S. Geological Survey Bulletin 1532. Washington, D.C.: Government Printing Office, 1982.

———. "Social Consciousness and World Maps." *Christian Century* 105 (1988): 190–92.

———. "The Space Oblique Mercator Projection." *Photogrammetric Engineering and Remote Sensing* 44 (1978): 585–96.

———. *The Space Oblique Mercator Projection: Mathematical Development.* U.S. Geological Survey Bulletin 1518. Washington, D.C.: Government Printing Office, 1981.

Snyder, John P., and Harry Steward. *Bibliography of Map Projections.* U.S. Geological Survey Bulletin 1856. Washington, D.C.: Government Printing Office, 1988.

Snyder, John P., and Philip M. Voxland. *An Album of Map Projections.* U.S. Geological Survey Professional Paper 1453. Washington, D.C.: Government Printing Office, 1989.

Sobel, Dava. *Longitude: The True Story of a Lone Genius Who Solved the Greatest Scientific Problems of His Time.* New York: Walker, 1995.

Sowell, Thomas. "Drawing a Line in Mercator's Defense." *Washington Times,* December 26, 1990.

———. *The Vision of the Anointed: Self-Congratulation as a Basis for Social Policy.* New York: Basic Books, 1995.

Stachurski, Richard J. "History of American Projections: The American Projection." *Professional Surveyor* 22 (April 2002): 16–22; (May 2002): 32–33.

Stage, Jeff. "Stamps Mark Olympics, WWII." *Syracuse Herald-American,* September 1, 1991, Stars sec., 29.

Stanley, Albert A. "Map Projections for Modern Charting." *Military Engineer* 40 (1948): 55–58.

Stephenson, F. Richard. "Chinese and Korean Star Maps and Catalogs." In *The History of Cartography,* vol. 2, bk. 2, *Cartography in the Traditional East Asian and Southeast Asian Societies,* edited by J. B. Harley and David Woodward, 511–78. Chicago: University of Chicago Press, 1994.

Stevenson, Edward Luther. *Portolan Charts: Their Origin and Characteristics, with a Descriptive List of Those Belonging to the Hispanic Society of America.* New York: Knickerbocker Press, 1911.

Stewart, John Q. "The Use and Abuse of Map Projections." *Geographical Review* 33 (1943): 589–604.

Stewart, John Q., and Newton L. Pierce. *Marine and Air Navigation.* Boston: Ginn and Company, 1944.

Stokes, Roy. *Esdailes' Manual of Bibliography,* edited by Stephen R. Almagno. Lanham, Md.: Scarecrow Press, 2001.

Stott, Peter H. "The UTM Grid Reference System." *IA, The Journal of the Society for Industrial Archeology* 3 (1977): 1–14.

Tannehill, I. R., and Edgar W. Woolard. "Gall's Projection for World Maps." *Monthly Weather Review* 64 (1936): 294–97.

Taylor, E. G. R. "Correspondance Mercatorienne [review]." *Geographical Journal* 127 (1961): 126–27.

———. *The Haven-Finding Art: A History of Navigation from Odysseus to Captain Cook.* London: Hollis and Carter, 1956.

———. "John Dee and the Map of North-East Asia." *Imago Mundi* 12 (1955): 103–6.

———. *The Mathematical Practitioners of Tudor and Stuart England.* Cambridge: Cambridge University Press for the Institute of Navigation, 1968.

Taylor, E. G. R., and M. W. Richey. *The Geometrical Seaman: A Book of Early Nautical Instruments.* London: Hollis and Carter for the Institute of Navigation, 1962.

Taylor, E. G. R., and D. H. Sadler. "The Doctrine of Nauticall Triangles Compendious." *Journal of the Institute of Navigation* 6 (1953): 131–47.

"Text and Translation of the Legends of the Original Chart of the World by Gerhard Mercator, Issued in 1569," *Hydrographic Review* 9 (November 1932): 7–45.

Times (London). Editorial, "Mercator Disciplined," January 28, 1954.

———. Obituary, "Arno Peters," December 10, 2002.

Tissot, M. A. *Mémoire sur la représentation des surfaces et les projections des cartes géographiques.* Paris, Gauthier-Villars, 1881.

Tobler, Waldo R. "Geographic Area and Map Projections." *Geographical Review* 53 (1963): 59–78.

———. "Medieval Distortions: The Projections of Ancient Maps." *Annals of the Association of American Geographers* 56 (1966): 351–60.

Torbert, John. "Maps and Mapmaking." *Bulletin of the American Geographical Society* 34 (1902): 197–210.

Tuchinsky, Philip M. *Mercator's World Map and the Calculus.* Lexington, Mass.: COMAP, 1978.

Turley, Gary. "Atlas Re-bound: Mercator's 1595 Atlas on CD-ROM." *Mercator's World* 6 (January/February 2001): 56–57.

Tyner, Judith A. "Interactions of Culture and Cartography." *History Teacher* 20 (1987): 455–64.

"Upside Down, Inside Out." *Economist* 293 (December 22, 1984): 19–24.

U.S. Army Map Service. "World Polyconic Grid." *A.M.S. Bulletins,* no. 1 (August 1943): 1–5.

U.S. Coast and Geodetic Survey. *Annual Report of the Director, United States Coast and Geodetic Survey to the Secretary of Commerce for the Fiscal Year Ended June 30, 1920.* Washington, D.C.: Government Printing Office, 1920.

———. *Annual Report of the Superintendent, United States Coast and Geodetic Survey to the Secretary of Commerce for the Fiscal Year Ended June 30, 1915.* Washington, D.C.: Government Printing Office, 1915.

———. "New Aeronautical Route Chart." *Military Engineer* 39 (1947): 178–79.

———. *Report of the Superintendent of the Coast and Geodetic Survey Showing the Progress of the Work from July 1, 1909, to June 30, 1910.* Washington, D.C.: Government Printing Office, 1911.

U.S. Department of Commerce, National Oceanic and Atmospheric Administration, National Ocean Service. *Nautical Chart User's Manual.* Washington, D.C.: Government Printing Office, 1997.

U.S. Department of the Army. *Grids and Grid References.* Technical Manual 5-241-1. Washington, D.C.: Headquarters, Dept. of the Army, 1967.

U.S. Department of Transportation, Federal Aviation Administration. *Pilot's Handbook of Aeronautical Knowledge,* rev. ed. Washington, D.C.: Government Printing Office, 1980.

———. *Report to Congress: National Plan of Integrated Airport Systems (2001–2005), August 28, 2002.* Washington, D.C.: N.p., 2002.

U.S. Library of Congress. "American Memory: Historical Collections for the National Digital Library." http://memory.loc.gov/.

U.S. Navy Department, Bureau of Navigation. *Projection Tables for the Use of the United States Navy: Comprising a New Table of Meridional Parts for the Mercator Projection with Reference to the Terrestrial Spheroid, and the Tables of the Polyconic Projection, as Used in the United States Coast Survey, Adapted to Areas Both of Small and Large Extent.* Washington, D.C.: Government Printing Office, 1869.

U.S. Navy Hydrographic Office. *American Practical Navigation: An Epitome of Navigation.* Washington, D.C.: Government Printing Office, 1962.

U.S. Ordnance Department. *Handbook of Artillery, including Mobile, Antiaircraft, Trench and Automotive Matériel,* Ordnance Department Document 2033. Washington, D.C.: Government Printing Office, 1925.

U.S. War Department. *Advanced Map and Aerial Photograph Reading.* Basic Field Manual FM 21-26. Washington, D.C.: Government Printing Office, 1941.

van Durme, M. *Correspondance Mercatorienne.* Antwerp: De Nederlandsche Boekhandel, 1959.

van Nouhuys, J. W. "Mercator's World Atlas 'Ad Usum Navigantium.'" *Hydrographic Review* 10 (November 1933): 237–41.

Varekamp, Thomas S. "A Discovery of Manuscript Maps by G. Mercator (1512–1594)." *Manuscripts* 22 (1970): 192–98.

"Views of the World: Maps in the British Prestige 'Press.'" *SUC* [Society of University Cartographers] *Bulletin* 33, no. 1 (1999): 1–14.

Vinge, Clarence L. "Mercator Projection: Its Basis in Elementary Mathematics." *School Science and Mathematics* 50 (1950): 394–401.

Vujakovic, Peter. "Arno Peters' Cult of the 'New Cartography': From Concept to

World Atlas." *SUC* [Society of University Cartographers] *Bulletin* 22, no. 2 (1989): 1–6.

———. "The Extent of Adoption of the Peters Projection by 'Third World' Organizations in the UK." *SUC* [Society of University Cartographers] *Bulletin* 21, no. 1 (1987): 11–15.

———. "Mapping the War Zone: Cartography, Geopolitics and Security Discourse in the UK Press." *Journalism Studies* 3 (2002): 187–202.

———. "Richard Edes Harrison and the Challenge to American Cartography." *Imago Mundi* 50 (1998): 174–88.

Wagner, Hermann. "Kartometrische Analyse der Weltkarte G. Mercators vom Jahre 1569." *Annalen der Hydrographie und Maritimen Meteorologie* 43 (1915): 377–94. Reprint, *Acta Cartographica* 25 (1977): 435–52.

Wagner, Karl Heinrich. "Das neue Kartenbild des Herrn Peters." *Kartographische Nachrichten* 23 (1973): 162–63.

Wallis, Helen M., and Arthur H. Robinson. *Cartographical Innovations: An International Handbook of Mapping Terms to 1900.* Tring, Herts, U.K.: Map Collector Publications, 1987.

Wall Street Journal. "Drawing the Line," June 8, 1989.

Warntz, William, and Peter Wolff. *Breakthroughs in Geography.* New York: New American Library, 1971.

Waters, D. W. *The Art of Navigation in England in Elizabethan and Early Stuart Times.* New Haven, Conn.: Yale University Press, 1958.

———. *Science and the Techniques of Navigation in the Renaissance.* National Maritime Museum Monographs and Reports 19. Basildon: Her Majesty's Stationery Office, 1976.

Wattenberg, Ben. "Real Shape of the World Changes." *Vicksburg* (Miss.) *Evening Post,* November 18, 1988.

Weber, Gustavus A. *The Coast and Geodetic Survey: Its History, Activities and Organization.* Baltimore: John Hopkins Press, 1923.

———. *The Hydrographic Office: Its History, Activities and Organization.* Baltimore: John Hopkins Press, 1926.

Wedemeyer, A. Letter dated August 18, 1931, in "Comments on the Existing Copies of the Mercator's Chart of 1569." *Hydrographic Review* 8 (November 1931): 204.

Weems, P. V. H. *Air Navigation,* 2nd ed. New York: McGraw-Hill, 1938.

———. *Marine Navigation.* New York: D. Van Nostrand, 1940.

Westfall, Richard S. "Mercator, Gerardus [Gerhard Kremer]." Catalog of the Scientific Community. http://es.rice.edu/ES/humsoc/Galileo/Catalog/catalog.html.

Whalley, Joyce I. *Writing Implements and Accessories from Roman Stylus to the Typewriter.* Detroit: Gale Research, 1975.

White, Denis, and others. "Comparing Area and Shape Distortion on Polyhedral-Based Recursive Partitions of the Sphere." *International Journal of Geographical Information Systems* 12 (1998): 805–27.

Whitfield, Peter. *The Charting of the Oceans: Ten Centuries of Maritime Maps.* Rhonert Park, Calif.: Pomegranate Artbooks, 1996.

———. *The Image of the World: 20 Centuries of World Maps.* San Francisco: Pomegranate Artbooks, 1994.

"Who Originated Mercator's Projection?" *Geographical Journal* 51 (1918): 270–71.

Wilford, John Noble. *The Mapmakers.* New York: Alfred A. Knopf, 2000.

Williams, Frances Leigh. *Matthew Fontaine Maury, Scientist of the Sea.* New Brunswick: Rutgers University Press, 1963.

Williams, J. E. D. *From Sails to Satellites: The Origin and Development of Navigational Science.* Oxford: Oxford University Press, 1992.

———. "Loxodromic Distances on the Terrestrial Spheroid." *Journal of Navigation* 3 (1950): 133–40.

Winchester, Simon. Review of *Mercator: the Man Who Mapped the Planet,* by Nicholas Crane. *New York Times,* January 23, 2003.

Wong, Frank Kuen Chun. "World Map Projections in the United States from 1940 to 1960." Master's thesis, Syracuse University, 1965.

Woodward, David. *The All-American Map: Wax Engraving and Its Influence on Cartography.* Chicago: University of Chicago Press, 1977.

Woolard, Edgar W. "Historical Note on Charts of the Distribution of Temperature, Pressure, and Winds over the Surface of the Earth." *Monthly Weather Review* 48 (1920): 408–11.

"A World-map by Hondius on Mercator's Projection." *Geographical Journal* 54 (1919): 123.

"The World Turned Upside Down." *Economist* 310 (March 25, 1989): 97.

Woytinsky, Wladimir S., and Emma Shadkhan Woytinsky. *World Population and Production: Trends and Outlook.* New York: Twentieth Century Fund, 1953.

Wright, Edward. *Certaine Errors in Navigation, Arising either of the Ordinarie Erroneous Making or Using of the Sea Chart, Compasse, Crosse Staffe, and Tables of Declination of the Sunne, and Fixed Starres Detected and Corrected.* London: Valentine Sims, 1599. Reprint (excerpts), *Hydrographic Review* 8 (May 1931): 84–100.

———. *Certaine Errors in Navigation Detected and Corrected,* 3rd ed. London: Joseph Moxon, 1657. Reprint, Early English Books, 1641–1700 [microfilm collection], 952: 27, Ann Arbor, Mich.: University Microfilms International.

Wroth, Lawerence C. *The Way of a Ship: An Essay on the Literature of Navigation Science.* Portland, Maine: Southworth-Anthoensen Press, 1937.

Young, A. E. *Some Investigations in the Theory of Map Projections.* London: Royal Geographic Society, 1920.

Zeller, Mary Claudia. *The Development of Trigonometry from Regiomontanus to Pitiscus.* Ann Arbor, Mich: Edwards Brothers, 1946.

Index